Fo

Introduction to the
Selection of Engineering Materials

To my loving wife, sons and daughter,
who have accompanied me round the world

Introduction to the
Selection of Engineering Materials

D.P. Hanley, *F.I.E.Aust.*

Senior Lecturer in Mechanical Engineering
Capricornia Institute of Advanced Education
Rockhampton, Queensland, Australia

VNR VAN NOSTRAND REINHOLD COMPANY
New York — Cincinnati — Toronto — London — Melbourne

**Published by Van Nostrand Reinhold Company Ltd.,
Molly Millars Lane, Wokingham, Berkshire, England**

*Published in 1980 by Van Nostrand Reinhold Company,
A Division of Litton Educational Publishing Inc.,
135 West 50th Street, New York, NY 10020, USA*

*Van Nostrand Reinhold Limited,
1410 Birchmount Road, Scarborough, Ontario, M1P 2E7,
Canada*

*Van Nostrand Reinhold Australia Pty. Limited,
17 Queen Street, Mitcham, Victoria 3132, Australia*

Library of Congress Cataloging in Publication Data

Hanley, D   P
   Introduction to the selection of engineering materials

   Bibliography: p.
   1. Materials.   I.   Title.
TA403.H364   620.1'1   79-27921
ISBN 0-442-30431-5

Printed and bound in Great Britain at
The Camelot Press Ltd, Southampton

Preface

An alternative title considered for this book was 'What I Would Like to Have Known About Materials Twenty Years Ago'. The author still remembers the uncertainty and sense of illiteracy in his first design job as a young engineer in having to specify materials to draftsmen. University training had provided a repertoire of '1020 and 4140' steels and 'aluminum alloy', although at the time—as a new graduate—no 'spec numbers' could be remembered for the latter!

The situation is still very prevalent today. After two decades of experience, most of which has been in industry, the author still finds that young and even mature engineers can often only recall the iron – carbon equilibrium diagram and never think beyond steel. The objective of this book is therefore to introduce at the undergraduate level at least the names of most materials of modern technology and their principal properties and applications. The third year engineering degree course from which this book has been developed follows a previous course which provides a materials science foundation.

Organization of the book largely follows that of Reinhold's excellent and annually published *Materials Selector*. Some rearrangement has been made to accommodate electrical engineering materials. The material in this book is based largely on student seminars given over a five-year period. Quantitative data have been kept intentionally to a minimum as these are readily available in references. The author acknowledges the many student contributions. If only one or two key features for each material are retained, the student will be well equipped in the future to know whether and where to look for further information.

Appreciation is expressed to the CIAE Council for their having allowed the writer to use part of his 1977 Study Leave to work on this book. Thanks are given to Dr. K.H. Sayers of CIAE and to the publishers for review and editorial assistances.

v

List of Tables

Contents

1
Overview of Materials Technology

This chapter presents some basic terminology, an overall classification system for engineering materials, and describes some fairly recent materials developments along with properties and applications. It is intended to serve as a framework of reference for the remainder of the text.

Basic Properties

Designers make use of a host of properties in specifying materials for structures and components. These include characteristics related to mechanical, thermal, chemical, electrical, and magnetic phenomena. Because structures carry mechanical loads—tension, compression, shear, bending, and combinations thereof—it is usual that mechanical properties be first considered. These include strength, stiffness, ductility, hardness, usually density, and resistance to elevated temperature. How are these properties measured? What are familiar values against which we may compare various materials? Concurrent with mechanical property assessment one must also consider factors such as cost, fabricability, and availability.

Of fundamental importance to the engineer in describing mechanical properties of any material is stress – strain behavior. The layman uses these words almost interchangeably—they have, however, specific meanings: stress is usually the cause (due to load); strain is the effect or response of material to stress (i.e., deformation). Figure 1 shows these definitions and some representative stress – strain curves for various materials.

On the left of Figure 1 is shown the basic nomenclature ... a section of material of area A is subjected to a load P. This produces a stress $\sigma = P/A$, which may be thought of as a tractive pressure and is measured in newtons per square millimetre (N/mm^2) or megapascals (MPa).

The element originally of length L increases by $\triangle L$ after load application and its extension or strain is designated ϵ—it is a dimensionless measure of elongation and is one of the parameters that quantifies ductility. Stiffness of the material is defined as stress divided by strain: where there is a linear relationship between the two quantities it is defined as E, Young's modulus; these parameters are most important quantities. They describe behavior and properties of all types of materials.

Consider the graphs of Figure 1. Firstly, for ferrous alloys—steels, of which

1

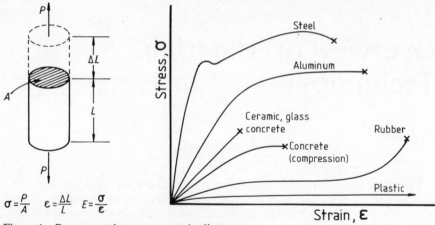

$$\sigma = \frac{P}{A} \quad \varepsilon = \frac{\Delta L}{L} \quad E = \frac{\sigma}{\varepsilon}$$

Figure 1　Representative stress – strain diagrams.

there are hundreds of compositions and heat treatments—the stress – strain curve is linear up to a point at which yielding occurs, followed by a flat region of plastic flow and then an increasing slope associated with strain- or work-hardening. Finally, maximum stress is reached, which is called the ultimate tensile strength of the material. Beyond this point, behavior becomes unstable and failure is imminent. With steels the stress at failure denoted by the termination point '×' is usually less than the ultimate or maximum stress.

Below the curve for steel is the curve for a typical aluminum alloy. Several different features in comparison to steel may be noted: its initial slope or modulus is less; its shape is quite different, being nonlinear up to failure; its strength is less; however, its elongation is greater. The stress – strain curve for concrete in compression is similar in shape to that of aluminum in tension. Clear timber and plywood when tested in compression and flexure also exhibit non-linear load – deflection curves.

Likewise, in terms of stress – strain behavior, one sees at a glance well-known characteristics of other familiar materials: ceramics, glass, and concrete in tension show a short, linear curve typical of low-strength, unyielding, or brittle materials; rubbers and elastomers show curves having low strength, high elongation, and an unusual stiffening characteristic as load increases and the polymer chains are straightened out. Finally, unreinforced plastics and polymers show curves which indicate low strength and high elongation compared to metals. Further, these materials exhibit flow characteristics which are implied by the flatness of their curves. It is thus concluded that materials can be classified according to their stress – strain response, which in turn derives from their compositions.

Classes of Materials

Materials may be divided into three classes:

1. metallic (including pure metals and alloys)

2

2. intermediate
3. nonmetallic

Recent developments in each of these classes will be touched upon in the following sections. In Class 1, new metallic materials and their forms are discussed in comparison to conventional ones. Ceramics, cermets, whiskers, and semiconductors will be covered in Class 2. Under nonmetallic materials, Class 3, developments in composites, plastics, and rubbers will be highlighted. There have been some remarkable breakthroughs in the past $10 - 15$ years in nonmetallics.

Materials in Class 1 are considered generally to have good strength, ease of working, moderate corrosion resistance, and low-to-moderate cost. Class 2 materials are primarily suited to compression loads in structures or to electrical, thermal, or magnetic applications because of their special properties. Whiskers, however, are very high strength materials. Composite materials in Class 3 possess outstanding strength, stiffness, and fatigue characteristics combined with very low density and excellent corrosion resistance. At present these are moderate to high in cost.

Class 1: Metallic Materials

Metals are next considered in greater detail. Alloys comprising two or more metals are discussed rather than pure metals, which have no real structural application. Table 1 gives approximate comparative values for several basic properties of some steels and aluminum, the latter being a fairly expensive aircraft alloy. Properties include strength, modulus, density, melting point, and cost.

Table 1   Major Structural Alloys

		σ (MPa)	E (GPa)	ρ (kg/m^3)	MP (°C)	Cost ($/kg)
Common	Steels					
	Structural	520	210	7800	1600	0.22
	Stainless	690	210	7800	1600	
	Maraging	2100	210	7800	1600	High
	UHS (ausformed)	3400+	210	7800	1600	
Not so common	Aluminum	520	70	2800	560	1.10
	Titanium (6Al – 4V)	1200	120	4500	1700	25
	Magnesium (Mg – Al)	340	40	1800	670	4
	Beryllium (Be – Al)	690	290	1900	1300	1700

It is seen in Table 1 that structural steel and aluminum are equal in strength. However, both the stiffness and density of aluminum are one-third those of steel. Aluminum thus offers a threefold strength-to-density advantage over structural steel, although its elevated temperature resistance is not as good. (Aluminum is rarely used at temperatures above 100 °C.) The lower stiffness

3

along with the fivefold cost factor is the main reason why aluminum is not used extensively in buildings, bridges, and so on, whereas it is used in, say, aircraft, where structural weight is of paramount importance. Its cost, however, is not excessive for demanding applications.

There are other higher-strength forms of steel (see Table 1), such as stainless steel—which in itself covers dozens of different alloys. Fairly recent developments include 'maraging' steels, which involve precipitation age hardening of martensite, a constituent phase of many steels. Maraging steels are high nickel content alloys containing low carbon and varying amounts of cobalt, molybdenum, titanium, and aluminum. Even higher in strength are the very high or ultrahigh strength steels—so called 'ausformed steels'. These steels derive their name from mechanical and thermal treatments while in the austenitic phase. Ausformed steels have strengths exceeding 3500 MPa, approaching seven times that of conventional steels! Whereas many high strength steels behave in a brittle fashion, there are also new developments with 'TRIP' steels which greatly increase ductility while preserving high strength. TRIP steels derive their name from 'Transformation Induced Plasticity' which involves high temperature straining during the phase change processes.

There are many other developments in steels aimed toward materials and fabrication economy. These include powder metallurgy and forging techniques. There are also new 'weathering' steels and many protection schemes which increase service life.

What are some of the competitors to conventional structural alloys—steel and aluminum? Several are shown in Table 1: titanium, magnesium, and beryllium alloys. An outstanding feature of titanium (usually alloyed with aluminum and vanadium) is that it has comparable strength to steel, a modulus intermediate between steel and aluminum, and a density half that of steel. Because of increased stiffness over aluminum and good creep resistance at elevated temperature, titanium is used in supersonic aircraft and accounts for 10 per cent of the Boeing 747 aircraft's structural weight. The latter usage amounts to about 40 tonnes per aircraft. Aerodynamic and engine heating in supersonic flight can produce skin temperatures as high as 300 °C, about three times the allowable temperature of aluminum! Use in supersonic aircraft is the first major engineering application of titanium. With presently available aircraft power plants, supersonic flight would not be possible without this metal.

The importance of stiffness or modulus in aircraft cannot be over-emphasized. Modulus dictates design for wing droop, lift loads, fuselage stiffness, landing loads, and many aspects of dynamics, vibrations, and fatigue. The price paid for the basic material—titanium—is roughly $25/kg at the time of writing (1979), 100 times greater than steel and 20 times that of aluminum.

It is also seen in Table 1 that magnesium (usually alloyed with aluminum, zinc, and rare earths such as yttrium and gadolinium), is not a particularly strong material and that it has a very low modulus (60% of that of aluminum). It is nevertheless the lightest known metal alloy. Its density is 40% less than that of aluminum. In many aircraft applications, minimum gage/minimum weight material is required for aerodynamic fairings only ... examples being secondary structures such as ducting, elevons, control surfaces, etc. With adequate stiffening, magnesium therefore finds some substantial use in noncritical lightweight

4

wrought structures. Greater use, however, is made of the metal in lightweight cast forms. There are some problems with sea water and salt air corrosion with magnesium alloys.

Shown last in Table 1 is beryllium. It is usually alloyed with aluminum. This is a high strength, very lightweight material and has the highest known stiffness of structural metals, one-third greater than steel and more than four times that of aluminum. At a cost, though, of approximately $1700/kg, or $1.7 million/tonne at the time of writing (1979), it is obviously used only in extremely demanding applications such as space vehicle structures, examples being the Agena engine nozzle support structure and the Lunar Excursion Module.

Considerable effort has been made and is still underway in reducing costs of titanium and beryllium and in solving difficult technical problems dealing with fabrication, stress corrosion cracking, and the toxicity effects (known as beryllicosis) associated with the latter.

Beyond the temperature capabilities of conventional and high performance metallic alloys are the high temperature metallics known as 'superalloys' and 'refractories' shown in Table 2.

Table 2   High Temperature Materials

	σ (MPa)	E (GPa)	ρ (kg/m^3)	MP (°C)
Superalloys				
(low Fe with/Ni, Co, Cr, Mo)	1050	210	7800	< 1650
Refractories				
(Cb, Mo, Ta, W)	1400	to 420	18200	2750
Carbon/graphite	35	14	1600	∿3600

The superalloys are low iron compositions with nickel, cobalt, chromium, and molybdenum. They are high strength alloys similar in modulus and density to steels; their permissible service temperatures, however, are much higher, exceeding 1000 °C in contrast to the 500 – 800 °C limit for steels. They find significant applications in engines, gas turbines, and power generation machinery. They are expensive and require costly fabrication procedures such as close tolerance hot forging.

The traditional refractory metals are columbium (also known as niobium), molybdenum, tantalum, and tungsten. These are high strength materials, some having moduli and densities twice those of steels. Although their melting temperatures are very high (∿2750 °C), most of these metals oxidize rapidly at very modest temperatures, which curtails their high temperature usefulness. A large volume of research has been applied to the problem of developing protective coatings for refractories which will adhere well under service conditions. Silicide coatings applied by diffusion processes are showing promise, but reliability is at present questionable.

A fairly recent development for high temperature metals has been 'TD' or thoria-dispersed nickel. Thoria acts to stabilize the pure metal and oxidation is not a problem. Of particular note is the fact that this material may be economically electroformed into a variety of complex shapes.

The only other uncoated metal which competes with nickel in volume applications is chromium. Its advantages are high strength and mechanical properties combined with thermal shock resistance. There are deterrents, however, which involve lack of room temperature ductility combined with high notch sensitivity, both factors being aggravated by nitridation by air at operating temperatures. Much work is being done on these problems for atomic energy and jet engine applications.

The 'near ultimate' material with regard to high temperature capability is shown last in Table 2—carbon or graphite, having a sublimation point about 3600 °C. Only hafnium carbide, a ceramic, surpasses this temperature level, and then only by several hundred degrees Celsius. Carbon or graphite in bulk form is neither a strong nor a stiff material, although it is very light. Of note is its use in refractory or ablation applications, particularly under chemically reducing environments—an example being rocket nozzle throat inserts. Oxidation is otherwise very severe and problems exist with coatings on carbon which are even more difficult than with metals.

Present significant uses of carbon and graphite are in electrodes for electric arc steel making and in chemical and rocket engine applications. A fairly recent development in forming the material is that of vapor deposition on to a substrate. This produces pyrolytic graphite, which is very pure, in shapes and sizes otherwise unfabricable.

Class 2: Intermediate Materials

Conventional uses of ceramics are such as to utilize their heat resistance, high hardness, compression strength, corrosion resistance, and certain nuclear, magnetic, and electrical properties. Well-known examples include bricks, roof tiles, furnace linings, electrical insulators, and magnetic ferrites—the last-mentioned being of tremendous importance in computer memories. Interest in ceramics insofar as wider engineering usage is concerned has been hampered by their notable weaknesses of brittleness, poor thermal shock resistance, and general unreliability under tensile stress. Work with ceramics in composites and in cermets has enabled production of some improved materials superior in crack propogation resistance. Some successes with fully dense alumina (aluminum oxide) and magnesia (magnesium oxide) have been achieved in increasing room temperature ductility.

Cermets are sintered or hot pressed materials consisting of combinations of ceramics and metals. Cermets combine the ductility and thermal shock resistance of metals with the desirable refractory qualities of ceramics. The metal serves as a binder or matrix for ceramic particles and increases thermal conductivity and shock resistance. This latter property imparts high temperature strength nearly equal to that of the pure ceramic. Metals such as iron, chromium, nickel, cobalt, aluminum, and molybdenum have been used with many types of carbides, borides, and oxides. These materials find use in turbine blades, cutting tools for hard metals, spinning tools, high temperature bearings,

6

and heavy current electrical contacts. Recent American work has evaluated boron carbide for hypersonic aircraft leading edges where loads and the environment are well beyond conventional materials capabilities.

Another interesting development is associated with glass ceramics known as 'pyrocerams'. These are materials which make use of a devitrification/heat treatment process for glass which results in strength values several times that of industrial glass, combined with a very low-to-zero thermal expansion coefficient. The attendant improvement in thermal shock resistance makes the material suitable for domestic cookware, laboratory bench tops, some heat exchangers, and in radomes which must be transparent to electromagnetic radiation.

Whisker technology involves growing single-crystals of very fine diameter fibers, usually of the intermediate materials class. Such fibers, although not continuous, have strengths in excess of 7000 MPa with moduli up to nearly 700 GPa. These outstanding physical properties, which result from the nearly perfect crystalline structure, may be used in reinforcing plastics, ceramics, or metals. Whiskers are available in alumina, sapphire, silicon carbide, and many other materials. Difficulties exist in the fabrication of parts and in fiber strength translation efficiency due to handling and orientation of whiskers. Wetting and interface bonding are also problems.

Semiconductors have revolutionized the field of electronics. Certain elements and combinations thereof form unusual electrical characteristics because of their atomic lattice structures. With just a mention of 'hole theory' and the 'avalanche effect', suffice it to say there are numerous important applications which make use of this property—namely, rectifiers, photocells, pressure gages, magnetometers, transisters, and energy convertors. Commonly used semiconductor materials are silicon and germanium. Many others such as antimonides, arsenides, and the tellurides of aluminum, bismuth, and gallium are used under high temperature, high power level conditions.

Thermistors are devices made of semiconductors (mostly oxides of several metals combined) which display high sensitivity to temperature differences as small as a millionth of a degree Celsius. These devices find use in precise temperature control and measurement and have allowed considerable miniaturization over conventional thermocouple techniques.

Class 3: Nonmetallic Materials

The nonmetallics comprise composites, plastics, and elastomers and wood.

Composites are heterogeneous two-phase materials consisting of a reinforcement (particles, flakes, and short, medium, or long fibers) surrounded by a protective load distributive matrix or binder. Matrix materials may be plastic, ceramic, or metallic. Reinforcements may be of various types such as glass, boron, graphite, asbestos, etc. In combination, the fiber/matrix system provides a structural or insulating material.

Composites are used because of their greater stiffness and/or strength along with lower densities than those of conventional alloys. Superior fatigue resistance is achieved because of the two-phase nature of the material and because of higher internal damping or energy absorption capacity. Other features are that fabrication is relatively straightforward and plant investment is often less than with some metal working plants, processes akin to plywood or glass fiber lamination being employed. A key point with composites is that of 'tailoring' the material and design. For instance, the choice of a particular fiber, a particular resin, a particular amount of resin, a particular number of plys, and particular ply orientations are design freedoms not possible with metals—i.e. putting the strength or stiffness where it is needed, and usually in a single fabrication step! Lastly, corrosion resistance of these materials is generally very good. Familiar examples are glass fiber/resin boat hulls, external automobile parts, etc. The major disadvantages with high performance composites are their costs (roughly $200 – 600/kg at the time of writing), the fact that sophisticated design techniques are required, and the fact that completely reliable quality control and inspection procedures are still being developed.

The word 'plastics' covers hundreds if not thousands of materials. Billions of tonnes are produced in the world each year and at a growing rate.

Plastics have come a long way since their principal role was that of low-cost substitutes for other materials. Many present-day materials have contributed to major developments in a variety of fields. Several high performance plastics are responsible for critical applications involving electronics, communications, and aerospace systems. Without the electrical properties, toughness, and extreme heat resistance of materials such as the new polyimides (withstanding temperatures to 540 °C) and high temperature nylon fabrics, reliability of such equipment would require bulky construction that could not be tolerated.

Another field in which plastics fill important needs is in medical equipment and prosthetics. From commonplace dental appliances, splints, disposable utensils, and hypodermic syringes to precision heart valves, bone adhesives, and joint implants, plastics have contributed greatly, not only in making life more comfortable but also in preserving it.

Plastics also perfom in applications where they do an equal or better job than other materials at lower costs. Many of these are apparent in automobiles. Today's average car is estimated to contain about 50 kg of plastics. Plastics figure prominently in safety features: headrests, crushable instruments, and energy-absorbing steering columns. They also make up almost 90% of the visible interiors: door panels, headliners, instrument clusters, upholstery, steering wheels, and electroplated plastic control knobs. Other uses are in exterior parts: energy-absorbing urethane foam bumpers, electroplated radiator grills, polypropylene fender fill plates, and airscoops are examples.

Other industries which make great use of plastics are those such as construction, packaging, appliances, aircraft, furniture, and agriculture. What are these materials, how are they made, and how are they classified?

Plastics are synthetic organic materials, resins, which are solid in finished form but, at some stage in their processing, are fluid enough to be shaped by application of heat and pressure.

In finished form, plastics consist of long-chain molecules called polymers. Smaller building-block molecules, which may be combined into polymers by catalysis, heat, and pressure, are known as monomers. Cross-linking of two or more polymers, a process in a sense analogous to alloying in metals, is known as copolymerization. Not all polymers are plastics—elastomers and paper are also polymers.

Plastics have been used to solve many engineering problems which could not be solved by use of other materials: self-lubricated and grit-resistant bearings, tubing which is flexible at cryogenic temperatures, lightweight high strength films, and large structures requiring little tooling investment.

There is a major drawback to synthetic resins: information on the newer plastics often cannot be disseminated fast enough to permit the average designer to evaluate and specify them as easily as he does metals. In some cases, property data are not available. Generally more effort is expended in selecting a plastic than for a metal.

Plastics have good combinations of properties, rather than extremes of any single property. For example, no unreinforced plastic approaches the steels in strength and stiffness. No plastic is as light in solid form as most woods, as elastic as soft rubbers, as scratch resistant or transparent as glass. Yet plastics are the only materials which can be simultaneously strong, light, flexible, and transparent.

From a classification standpoint, plastics may be divided into five main types:

1. low friction
2. structural
3. chemical and thermal
4. electrostructural
5. light transmission

Properties required with Type 1 are as follows: a low coefficient of friction, even when nonlubricated, along with high abrasion resistance; fair to good form stability; heat resistance; and corrosion resistance. Suitable plastics of this type are filled and unfilled fluorocarbons (Teflon), nylons, acetals, and high density polyethylenes. These plastics generally compete with babbitts, bronzes, cast irons, prelubricated woods, graphite, and some cermets.

Properties required of Type 2 plastics are high tensile and high impact strengths. Good fatigue resistance and stability at elevated temperatures are needed along with machinability and moldability to close tolerances. Suitable plastics are nylon, TFE-filled acetals, polycarbonates, and fabric-filled phenolics and epoxies. Compared with cast iron, brasses, and steels, these offer advantages with respect to weight reduction and resistance to abrasive or corrosive environments and fatigue.

Properties required with Type 3 plastics are resistance to temperature extremes and to a wide range of chemicals. Minimum moisture absorption and fair-to-good strengths are also needed. Suitable plastics are fluorocarbons, chlorinated polyether, polyvinylidene fluoride, polypropylene, high-density polyethylene, and epoxy-glass. Metals competing with Type 3 plastics are stainless steels, titanium, columbium, and other premium metals.

The principal property required with Type 4 electrostructural plastics is excellent electrical resistance at low-to-medium frequencies. High strength and impact properties, good fatigue and heat resistance, and dimensional stability at elevated temperatures are also needed. In this group are allylics, alkyds, aminos, epoxies, phenolics, polycarbonates, polyesters, polyphenylene oxides, and silicones. The Type 4 plastics compete with ceramics and glasses.

Good light transmission in transparent or translucent colors is required for the Type 5 plastics. Formability, moldability, and shatter resistance are important requisites. Major plastics of this type are the acrylics, polystyrenes, cellulose acetates and butyrates, ionomers, rigid vinyls, polycarbonates, and medium impact styrenes. These find many applications where glasses are not suitable.

Rubber classifications will be dealt with later. For the present, however, it may be noted that numerous problems exist with such natural materials ... strength deficiencies, flex cracking, ageing, low tear strength, poor abrasion resistance, poor thermal properties, and variable (usually poor) corrosion resistance. There have been developed over recent years, however, many synthetic rubbers which may be molded and extruded and which possess superior properties. These include fluorosilicones and fluorocarbons, polysulfides and acrylates, urethanes, chloroprenes, and ethylene propylene. In contrast to natural rubber, which is limited in application to temperatures of -50 to $+80$ °C, some of these materials can be used from -100 °C to almost 260 °C, and in the presence of very hostile environments due to weather, ozone, radiation, and chemicals.

Applications of synthetic rubbers include footwear, power belting, molded mechanical and electrical goods, sports equipment, tires, gaskets, O-rings, chemical tank linings, pharmaceutical products, and many others which would not exist without materials technology.

2
Ferrous Alloys

This chapter emphasizes factors in the selection of ferrous alloys. There are two basic considerations in the selection process:

1. general ones involving function, application, cost, and availability;
2. specific ones concerned with the following detailed engineering properties:

Time/temperature-dependent strengths	Machinability
Modulus	Formability
Fatigue	Weldability
Notch or fracture toughness	Creep/relaxation
Abrasion resistance/hardness	Heat treatment
Corrosion resistance	Damping
	Magnetic, electrical, and nuclear properties

Classification

Ferrous alloys comprise irons, cast irons, and steels—the most widely used structural material. They are comprised of iron and carbon and other elements, the last-mentioned either intentionally added or present in trace amounts. Ferrous alloys are broadly classed according to carbon content expressed in weight percentage as below:

$$0 - 0.015\% \quad C \; - \; \text{irons}$$
$$0.01 - 2\% \qquad C \; - \; \text{steels}$$
$$2 + \; - 5\% \qquad C \; - \; \text{cast irons}$$

Alloys with carbon contents $> 5\%$ have no significant usage because of their poor properties.

How does one go about selecting a ferrous alloy? Certainly not because of an exotic or appealing trade name nor because of a minor alloying element. Why not? Because the properties of practically all steels are virtually interchangeable according to heat treatments or cold working.

Why are other elements added in ferrous alloys? Obviously to achieve specific chemical and/or metallurgical properties. There are nine major alloy series as designated by the Society of Automotive Engineers (SAE). The series were derived from the American Iron and Steel Institute (AISI) and the series is employed in Australian Standards.

Major references to alloy specification are given in the American Society of Metals (ASM) and the Australian Institute of Metals (AIM) Handbooks, British Standard BS 970, the UK (Morgan-Grampian) and US (Reinhold) Materials Selectors and in sources such as Machine Design (Penton) Reference Issues.

The SAE 1000 series steels are designated plain carbon. These account for more than 90 percent of the total steel production. The role of carbon as a solid solution in iron is to increase strength, hardness, and wear resistance. However, as the carbon content is increased, ductility and weldability are decreased. Typical designations in the plain carbon series are

10XX – plain machining grade
11XX – resulfurized
13XX – with 1.75% Mn

where XX generally indicates carbon content in hundreths of a percent. All steels may contain up to 0.5% Mn even though manganese is not specified in the composition.

High manganese steels do not come under the SAE system, but they are very tough, wear resistant, difficult to machine, and fairly low in cost. Steel containing 11 – 14% Mn and 1.0 – 1.3% C is known as Hadfield's steel and is similar in characteristics to some stainless steels.

The 2000 series of nickel steels have largely been replaced by the 5000 chromium alloy series. Nickel has in past years come mainly from Canada. Nickel mining in North Queensland and in Western Australia could possibly alter the supply of the metal. The functions of nickel when alloyed in steel are to increase tensile strength without reducing elongation appreciably and without reduction in area, to decrease heat treatment temperatures, to increase quench time and thereby minimize thermal distortion and possible cracking, and to increase impact strength—particularly at low temperatures. Nickel steels may be carburized. They are fairly expensive. There are four major groupings of nickel steels:

1. <6% Ni   Structural: with 0.1 – 0.5% C and ≤ 0.8% Mn.
              Case hardening: with 0.1 – 0.25% C.
2. 20 – 30% Ni   Nonmagnetic, tough, low coefficients of
                 thermal expansion.
                 Used in engines and turbine blades.
                 Contain 1.2 – 1.4% Mn and 0.5% Cr.
3. 30 – 40% Ni   Extremely low coefficients of thermal
                 expansion.
4. > 50% Ni   Very high magnetic permeability.

Series 3000 alloys contain nickel and chromium, the latter obtained from South Africa, Western Australia, and Turkey. These alloys have high strength and high hardenability. They are very wear resistant and are used in tooling and bearing races. The Ni – Cr alloys form the basis of stainless steels and offer a wide range of properties. Typical low nickel compositions are:

12

0.1 – 0.5% C
1.0 – 4.5% Ni
0.45 – 1.75% Cr
0.3 – 0.8% Mn

Some of the alloys are prone to temper brittleness and require special heat treatments. With carbon contents < 0.35%, they can be case hardened.

Stainless steels of higher nickel content than the above are divided into three types:

1. martensitic – containing 10 – 14% Cr
2. ferritic – containing 14 – 18 or 23 – 30% Cr
3. austenitic – containing 15 – 26% Cr + 7 – 20% Ni

A common stainless steel known as '18 – 8' is of type 3 and contains 18% Cr and 8% Ni. It is weldable, nonmagnetic, and can be hardened by cold working; it can also be annealed. A fairly common problem with high chromium alloys is that of carbide precipitation at grain boundaries, which weakens the material and promotes corrosion effects. High chromium alloys containing up to 3% Si provide better heat resistance.

The 4000 series alloys are chromium – molybdenum types. Molybdenum comes primarily from Colorado. In combination chromium and molybdenum provide good ductility, toughness, and machinability, along with excellent hardenability. A few of the more common types are:

40XX with 0.25% Mo
41XX with 0.80% Cr + 0.16% Mo
43XX with 1.82% Ni + 0.65% Cr + 0.25% Mo
46XX with 1.70% Ni + 0.22% Mo
48XX with 3.50% Ni + 0.25% Mo

Chromium steels in series 5000 contain up to 1% Cr for similar reasons as with the series 3000. Chromium may be used as a principal element or in a combination with others such as 0.2 – 1.7% Cr with 0.15 – 1.1% C and Mn < 1.5%. Hardenability and quenchability are main features of steels containing chromium. Low chromium alloys are usually 'structurals' whereas high chromium alloys are used for acid and corrosion resistance.

The 6000 series alloys contain chromium and vanadium, the latter being obtained from the USA and the Carribean. These alloys are generally resistant to grain growth and vanadium serves to accentuate other properties, particularly creep and fatigue resistance. Some series 6000 alloys are used in heavy duty springs and axles.

Series 7000 alloys contain up to about 6% tungsten. They are known for excellent hardness and wear resistance and are much used in tool and die applications.

The metals in series 8000 are known as 'triple alloys' since they contain Ni, Cr, and Mo. They are especially suited to elevated temperature applications and hostile environments. The presence of molybdenum serves to reduce temper brittleness. Manganese may also be added, partly replacing the nickel content without appreciable effect on properties, thereby reducing cost.

13

The 9000 series silicomanganese steels (1.72 – 2.25% Si and 0.6 – 0.9% Mn) are very important in electrical applications. Silicon enhances electrical resistivity and magnetic permeability. These alloys are often used in transformer cores and in heavy duty leaf and coil springs. Chromium – vanadium steels are also used in the latter applications. Vanadium is a strong deoxidizer and particularly improves fatigue strength. Tensile strength is good; however, chromium – vanadium steels tend to exhibit brittle behavior.

Last in the SAE designation scheme are specially classified boron steels indicated, for example, by 51BXX, 81BXX, or 94BXX. Hardenability is greatly increased with as little as 0.001% B in Cr – Ni – Mo steels.

A useful chart for the selection of some common ferrous alloys is given in Table 3.

Specification

The Australian specification system for alloy designation is illustrated in Figure 2. Steels may also be specified as in British Standards as later described according to chemical composition, hardenability requirements, or mechanical properties.

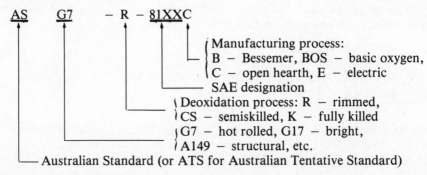

Figure 2   The Australian specification system for alloy designation.

Four common constructional Australian steels—all of similar mechanical properties—are designated AS A149, 157, 81, and 92. These contain 0.2 – 0.4% C, 0.5 – 0.85% Mn, 0.05% (max.) P and S. AS A135 and A151 are higher yield strength alloys containing greater amounts of Mn. AUSTEN 50 is the strongest available and contains Si, Ni, Cr, and Cu. Its Charpy strength is about 50% greater than the other constructionals.

Some typical applications of commonly used steels are given below:

Aircraft
- Sheet, tubing, gears, engine frames – 4130, 6352-62, 8630
- Cylinder barrels – 4140
- Cranks, propeller shafting – 3140, 3310

Automobiles, light trucks, tractors – 1340, 3130, 4047, 8640
Heavy trucks, diesel cranks – 5046
Valves – 3140, 4140, 8740

Table 3 Ferrous Alloy Selection Chart*

	Group 1										Group 2											Group 3		
	C-1018	C-1212 / B-1112	C-1213 / B-1113	Ledloy	C-1117	4615	4620	8620	E-3310[1]	Nitralloy 135 Mod[2]	E-4130[1]	E-8630[1]	C-1045	C-1141	8640[1] / 8740[1]	4140[1]	E-4340[1]	4150[1]	4150 modified[1]	E-6150[1]	C-1144 tempered	4140[3]	4340[3]	4150
Determine Group																								
1. Case-harden?	Yes	Yes	Yes	Yes	Yes	Yes	Yes	Yes	Yes	Yes														
2. Through-harden?											Yes	Yes	Yes	Yes	Yes	Yes	Yes	Yes	Yes	Yes	Yes	Yes	Yes	Yes
3. Practical pre-heat treated stock?									Yes	Yes					Yes	Yes	Yes	Yes	Yes	Yes		Yes	Yes	Yes
4. Directly from hot or cold formed stock?	OK	OK	OK	OK	OK								OK	OK							OK			
Determine steel																								
5. Hardness[4]	20 / 64	12 / 62	12 / 62	12 / 62	20 / 64	30 / 64	35 / 64	35 / 64	45 / 62	48 / 72	48 / 52	48 / 52	48 / 60	48 / 58	54 / 58	54 / 58	55 / 60	60 / 62	60 / 62	60 / 62	23 / 62	28 / 60	28 / 60	34 / 62
6. Maximum cross-sectional area?[5]	480	320	320	320	480	970	1130	1130	2260	2580	1770	1770	1300	1450	2420	2900	4190	3550	3550	3220	1940	3060	4190	3550
7. Induction or flame hardening required?		OK	OK	OK	?	?	?	?			OK	OK	OK	OK	OK	OK	OK	OK	OK	OK	OK	OK	OK	OK
8. Welding required?	OK	OK	OK	OK	?	OK	OK	OK			OK	OK												
9. Machinability required?[6]	55	100	130	160	90	55	52	50	35	35	67	65	55	70	60	60	50	55	70	55	70	45	40	65
10. Cost factor [7,8] HR	0.78	1.00	–	–	0.88	1.26	1.25	1.25	1.71	2.00	–	–	0.84	0.91	1.28	1.12	1.75	1.23	1.27	1.66	–	1.47	1.71	1.67
10. Cost factor [7,8] CF	0.95	1.00	1.02	1.12	0.97	1.51	1.49	1.37	–	2.28	1.81	1.84	0.95	1.00	1.71	1.63	2.05	1.63	1.69	–	1.23	1.84	–	1.91

* After HAUPTLY, Materials Engineering, 1955.

(1) Annealed.
(2) Heat treat before nitriding.
(3) Heat treated.
(4) Rockwell 'C'.
(5) In square millimetres.
(6) Arbitrary values given: those >100 infer easier machinability; <100 more difficult machinability.
(7) HR = hot rolled. CF = cold finished.
(8) Cost values >1 infer greater relative cost; <1 infer lesser relative cost.

15

Chemical industry	$\{$ high temperature	$\{$ Mo $-$ Si
		Cr $-$ Mo
	$\{$ low temperature	Ni

BS 970 (revised 1970 – 72) replaces the previous European 'EN' system for wrought engineering steels. The new system makes use of a six-symbol alphanumeric code:

1. The first three digits designate the type of steel:

000 – 199	C and C – Mn types
200 – 240	free cutting
250	Si – Mn spring steels
300 – 499	stainless, heat resistant valve steels
500 – 999	alloy steels in groups of tens

2. The fourth symbol, a letter, has one of the four following meanings:

A   supplied to chemical composition—mainly used on bar and wire
H   supplied to hardenability requirements
M   supplied to mechanical property requirements— mainly used on mill products
S   stainless or heat resistant alloy

3. The fifth and sixth digits correspond to the 0.XX% carbon content.

There are many other standards for cast ferrous alloys given in the AIM Handbook. These apply to 'low' and 'high' alloys. Alloys are designated low if the following conditions hold:

1. Contents for the following elements exceed given values as shown below:

Al	0.3%	Ni	0.5%
Cr	0.5%	P	0.12%
Co	0.3%	Si	2.0%
Cu	0.4%	S	0.1%
Pb	0.1%	S + P	0.2%
Mn + Si	2.0%	W	0.3%
Mo	0.1%	V	0.1%

2. The phase diagram matches a comparable carbon steel.

If the reader has not recalled concepts of phase diagrams, time – temperature – transformation diagrams, and hardenability, he would do well to review them. The key to ferrous alloy selection is that of achieving the desired properties at the lowest practical cost.

Cast Irons

Cast irons may often be used in place of steels at considerable cost savings. There are five basic types:

1. unmodified—which includes gray, mottled, and white irons, which names derive from the characteristic color of fresh fractures;

and, four higher quality types:

2. malleable
3. spheroidal (or nodular) graphite
4. alloy
5. Meehanite

The last above type is a proprietary alloy. The properties of cast irons are generally dictated by the quantity and form of 'free graphite' present in the microstructure.

Design and production advantages of cast irons include:

low tooling and production costs
ready availability
good machinability without burring
readily cast in complex shapes
excellent wear resistance and high hardness—
    particularly white irons
high inherent damping

Properties of most cast irons are not appreciably affected at temperatures up to 340 °C and they can be used at temperatures above 480 °C. Typical compositions of unmodified cast irons are shown in Table 4.

Table 4   Typical Compositions of Unmodified Cast Irons

Type	C Graphite(%)	C Total(%)	Si(%)	Mn(%)	P(%)	S(%)
Gray	2.5 – 3.1	3.2 – 3.5	1.8 – 2.4	0.8 – 0.6	0.3 – 1.0	0.10
Mottled	1.5	3.3	0.9	0.4	0.2	0.13
White	0	3.3	0.5	0.5	0.5	0.15

The minor constituents in unmodified cast irons are S, Mn, and P. With low Mn content, S 'fixes' cementite and opposes the graphitizing effect of Si. With high Mn content, S forms MnS, which is an innocuous phase. With high Mn and S, Mn_3C is formed, which considerably hardens the alloy. The effect of P is to increase fluidity in casting. The compound Fe_3P which is formed is known as steadite. Machinability is very much decreased with P contents $> 0.25\%$.

Malleable irons may be worked to smooth the shape of free graphite particles and thereby reduce stress concentration effects. There are two types: blackheart—which uses a long-term anneal in a neutral atmosphere, and whiteheart—which is similar but employs an oxidizing atmosphere. Malleable iron was the forerunner of modern steel. It is strong, ductile, has good impact and fatigue properties, and can be readily forged, welded, soldered, brazed, and flame hardened. Corrosion resistance is generally better than for carbon and alloy steels, especially in salt air.

17

Spheroidal graphite irons achieve rounded free graphite in their microstructure through melt innoculation with Ca, Ce, or Mn. They are more expensive than ordinary gray irons and can be heat treated. Spheroidal or ductile iron does not have impact resistance as good as that of equivalent strength steels. High temperature resistance is achieved up to 815 °C.

Alloy cast irons have the following characteristics: reduced impurity levels, improved control of graphite flake size, and improved mechanical properties. Chromium and molybdenum are frequently used alloying elements because they readily form carbides which promote hardening. Nickel added to gray iron forms the most important group of alloy cast irons. Nickel in various amounts in gray iron gives the following properties:

1. < 2%   – increased strength and toughness
2. 2 – 6%   – heat treatable irons
3. 10 – 40% – very hard, shock resistant martensitic irons
                 such as 'Nihard'

White iron containing 5% Ni is known as austenitic iron. It has exceptionally good corrosion resistance. 'Ni-Resist' are high nickel content cast irons used in chemical industries. Because of their nonmagnetic nature, they are also used in certain electrical applications. High alloy irons are extremely difficult to machine and are well suited to handle abrasive slurries.

Meehanite cast irons are formed in a proprietary process utilizing melt innoculation with $CaSi_2$. There are four types of these alloys: general use, abrasion resistant, heat resistant, and corrosion resistant.

3
Nonferrous Alloys

Nonferrous alloys of greatest importance are listed in Table 5 along with some generic characteristics.

Aluminum

It has been estimated that 8% of the Earth's crust is composed of aluminum, usually found in the oxide form and known as bauxite. Extensive bauxite mining is carried out in North Queensland and large-scale refining of it is carried out in Central Queensland. Outstanding properties of the metal and its alloys are:

low density
high strength-to-weight ratio
excellent corrosion resistance
high thermal and electrical conductivity
nonmagnetic and nonsparking
high reflectivity
readily worked and joined
nontoxic

Aluminum, copper, lead, and zinc are important nonferrous Australian natural resources.

Applications

Because of the world oil shortage, emphasis is being placed on the production of smaller and lighter vehicles for transport. Hence aluminum is finding increased usage in engine blocks and automobile bodies as well as in railroad systems. Traditional use has been (and is) in aircraft applications. The high corrosion resistance offered by aluminum alloys enables them to be used where maintenance costs must be minimized, such as in roofing materials and in sliding products such as door and window frames.

Because of its conductivity and nontoxicity, aluminum finds extensive use in cookware, medical equipment, and in the food processing and packaging industries. Since it is nonmagnetic, aluminum is used in shielding in electrical and instrumentation applications, while its nonsparking characteristic allows its use

Table 5 Some Generic Properties of Nonferrous Alloys

Alloy	Low density structurals	High electrical and thermal conductivity	Poor electrical and good thermal conductivity	Readily worked and joined	High corrosion resistance	High modulus	Low modulus	Used in brass and bronze alloys	Used in magnets	High density	High temperature structurals	Used in steel alloying	Cryogenic applications	Nuclear applications	Used in nonferrous alloys	Toxicity problems
Aluminum	✓	✓		✓	✓			✓							✓	
Beryllium	✓	✓			✓	✓		✓						✓	✓	✓
Cobalt		✓			✓				✓	✓	✓	✓		✓	✓	
Copper		✓		✓	✓			✓		✓					✓	
Columbium				✓	✓		✓			✓	✓		✓			
Lead				✓			✓	✓							✓	✓
Magnesium	✓	✓					✓								✓	
Molybdenum		✓				✓					✓	✓				
Nickel		✓			✓	✓		✓				✓				✓
Precious metals		✓			✓					✓					✓	
Rare earths			✓									✓			✓	
Refractory metals					✓	✓				✓	✓			✓		
Tin		✓		✓	✓		✓	✓							✓	
Titanium	✓				✓						✓	✓			✓	
Uranium										✓				✓		✓
Zinc		✓		✓	✓			✓							✓	
Zirconium					✓									✓		

in the presence of explosive mixtures. As aluminum reflects about 90% of visible light, it finds use in lighting fixtures and in building construction as an insulating medium. Artificial radioisotopes may also be produced from the metal. Almost all methods of forming and joining may be performed with aluminum. It can also be anodized, plated, electroformed, flame sprayed, and is widely used in castings and rolled specialty products.

Classification of Alloys

Alloy classification is according to the Aluminum Development Council of Australia system (AS 1664-1975) and is similar to that used in the U.S.
Principal wrought alloys are as follows:

1XXX. > 99% aluminum content. Referred to as 'dead soft' and is not-heat treatable. Second digit indicates modification in impurity limits. Last two digits indicate purity over 99%. The series is used primarily in electrical applications.

2XXX. Copper alloys. Second digit (as with the 3XXX series) designates the principal alloying element. The series is heat treatable and has mechanical properties similar to mild steel.

With precipitation hardening of alloys in this series, yield strength is increased, but corrosion resistance is somewhat decreased. A cladding of pure aluminum ('Alclad') is sometimes employed on these to improve corrosion resistance. Alloy 2024 is widely used in aircraft.

3XXX. Manganese alloys. Non-heat treatable. Although used infrequently, the series are general purpose alloys of moderate strength and good workability.

4XXX. Silicon alloys. Non-heat treatable. The alloys have low melting points and good ductility and are used in welding wire and in brazing.

5XXX. Magnesium alloys. Not-heat treatable, widely used alloys having moderate-to-high strength, hardness, and good corrosion resistance to marine environments.

6XXX. Silicon – magnesium alloys. Non-heat treatable, versatile alloys having good formability, corrosion resistance, and moderate strength.

7XXX. Zinc alloys. Heat treatable types having exceptional strength when alloyed with small amounts of magnesium. 7075-T6 is widely used in aircraft structures.

Cast aluminum alloys are designated by three digits. The first digit refers to the basic alloy series and the other digits refer to physical properties and forms available. Types are shown in Table 6.

There are also available a number of proprietary aluminum cast alloys which are comprised of various amounts of Cu – Mg – Zn, Zn – Cu, and Zn – Mg. Some of these have particularly good elevated temperature strength retention.

21

Table 6   **Types of Cast Aluminum Alloy**

Series	Major alloying elements	Features
100	Nearly pure form	Electrical applications— motor rotors, etc.
200	Copper	Increased strength, reduced shrinkage, improved machinability. Decreased casting fluidity and corrosion resistance.
300	Copper – silicon	Silicon improves casting while copper improves strength. Reduced ductility and corrosion resistance.
400	Silicon	Widely used. High initial fluidity and flow shrinkage.
500	Magnesium	Excellent mechanical properties, high corrosion resistance, good machinability. Casting difficult.
600	Silicon – Magnesium	Good corrosion resistance and strength.
700	Zinc	Good impact resistance. High zinc contents result in high shrinkage.
800	Tin	Anti-frictional applications such as bearings.

Beryllium

Beryllium might be classified as a precious metal as its abundance is less than 0.0005% of the Earth's composition. The principal source of the metal is beryllium – aluminum silicate, a mineral, which contains about 4% of the metal.

Outstanding properties of beryllium and its alloys are:

high strength/weight ratio
very high elastic modulus
high thermal/electrical properties
extremely brittle when unalloyed
resists oxidation in air to 820 °C
mostly acid resistant (except to dilute HNO_3); suffers nitridation at high temperatures
toxicity problems which are respiratory in nature, hence special precautions are required in machining
low neutron absorption
high permeability to X-rays

Applications

About 80% of beryllium produced is used in the alloy beryllium – bronze. The usual limit of content of beryllium in beryllium – bronze is 3%. Beryllium coppers are precipitation hardening alloys and may be heat treated and age hardened. Characteristics are high strength, high stiffness, high fatigue resistance, good formability, toughness, and corrosion and wear resistance. These alloys are used in springs, nonsparking tools, molds for plastics, and strong mechanical parts. Silicon may be added for improved strength and hardness while nickel may be added to refine grain size and decrease ductility.

A fairly recent development is 'Lockalloy', which is a beryllium – aluminum alloy used in airframes.

Beryllium may be added to nickel to produce alloys resembling high strength steels, comparable in properties and having excellent corrosion resistance. Such alloys are used in some turbine blades, precision springs, engine valves, and surgical instruments. Beryllium may also be added to iron or steel containing nickel and chromium to produce alloys having high elevated temperature strength and hardness.

Because of its very high cost, the pure metal is only used in demanding cases such as reflecting surfaces on space structures, some heat sink space applications, X-ray windows, and in high temperature electronic circuitry and some small instrument motor windings.

The metal and Be – 38 Al alloy in wrought form can only be worked slightly. Hot working is usually carried out from 540 to 1100 °C and annealing is carried out under vacuum. Pure beryllium machines like cast iron whereas the alloy machines like magnesium. The former is generally brazed with Zn, Al, or Ag alloys; the latter may be TIG or MIG welded or brazed as with aluminum alloys. Forms available include sheet, rod, bar, tube, wire, foil, extrusions, and powder. Major uses are in nuclear applications.

Cobalt

Cobalt is found in small quantities in the ores of cobalt, copper and iron sulfides, arsenides, and arsenosulfides. The element is only about one-quarter as abundant as nickel, cobalt, constituting about 0.002% of the Earth's crust. In pure form cobalt is silvery white in color with a bluish cast. Cobalt tends to exist as a mixture of two allotropes over a wide range of temperatures. At about 417 °C pure cobalt transforms from a hexagonal close-packed structure to face-centered cubic. Cobalt is magnetic, but unlike nickel and iron it retains its magnetism even at white heat. At 1150 °C cobalt loses all magnetic properties.

Applications of cobalt and its alloys include the following.

Magnets

Cobalt plays an important part in the production of Alnico-type magnets which have a composition of 14 – 30% Ni, 6 – 13% Al, and 5 – 35% Co, with the balance iron. The addition of cobalt greatly increases coercive force and residual magnetism.

Iron – cobalt alloys possess very high positive magnetostriction. An example of this type is Perminvar, an alloy having the highest value of magnetostriction,

23

which also maintains a constant permeability over a range of flux densities.

Cobalt oxide powder is also used in the production of magnetic materials for applications such as electric motor relays, small electric motors, generators, and switching gear.

Stellites

The use of cobalt as a base material produces alloys of superior hardness at both normal and high temperatures. Stellites are resistant to oxidation and to most chemicals including nitric acid, phosphorus acids, organic acids, and caustic alkalis. Because of their outstanding properties, stellites are often used in hardfacing processes.

Some stellite grades are listed in Table 7. Characteristics of the various grades are as follows:

Table 7   Some Grades of Stellites

Stellite grade	Chemical composition (%)					
	Ni	Co	Cr	C	W	Mo
3	–	52	30	2.4	13	–
4	–	53	31	1.0	14	–
7	–	66	26	0.4	6	–
8	–	63	30	0.2	–	6
X40	10	56	25	0.3	7	–

Stellite 3	Hardest
	Resistant to indentation at high temperatures
	Used in valve seat inserts for internal combustion engines
Stellite 4	Resists corrosion and abrasion
	Excellent hot strength
	Used in dies for extrusion of brass and copper
Stellites 7 and 8	High ductility
	Resistant to corrosion and thermal shock
	Basis of alloys for surgical implants
Stellite X40	High resistance to creep
	Used in gas turbines and stators

High Speed Steels

Cobalt is an essential constituent of a group of steels capable of heavy cutting at high speeds. Cobalt also serves as a cementing compound for tungsten carbide, a material having extreme hardness.

24

Cobalt 60

Cobalt 60 is a radioactive isotope which has industrial and medical uses such as for the examination of welds, tracing pipe blockages, internal medical examinations, and the sterilizing of equipment.

Other Uses

Cobalt oxides are used extensively in the ceramic industry, mainly in enabling enamels to adhere to metals. Cobalt linoleate and tungstate are used as driers for paints and varnish oils.

Copper

Copper occurs naturally with elements such as lead, nickel, silver, and zinc. It is widely used in industry both as a pure metal and as an alloying material.

In Australia, the largest known copper deposits are at Mt. Isa. Other major sources are near Tennant Creek and at Cobar. The large Mt. Morgan deposit is nearing exhaustion. In world production, the USA, Chile, Zimbabwe-Rhodesia, the USSR, Canada, Zambia, and Japan lead Australia.

Properties

Copper and copper base alloys have moderate strength and hardness, excellent corrosion resistance, and good workability, along with the following additional characteristics:

    pleasing, wide range of colors
    very high electrical and thermal conductivity
    nonmagnetic properties
    superior properties at subnormal temperatures
    ease of finishing by polishing and plating
    good to excellent machinability
    excellent resistance to fatigue, abrasion, and wear
    relative ease of joining by soldering, brazing, and welding
    moderate cost
    availability in a wide range of forms and tempers
    greater workability than ferrous materials

Production

Copper is found in small quantities in its ores. The ore is crushed and ground to a fine powder, and floatation is used to separate the copper-bearing grains. Various methods are used to refine copper, such as fire refining and electrolytic refining.

25

In fire refining, the impure copper is melted and oxidized to remove impurities. When the slag so produced has been removed, poles of green hardwood are thrust into the bath to remove the oxygen by combustion. After this is completed, the copper is poured into molds.

The electrolytic refining process is used for producing very pure copper. The process employs thick slabs of impure copper which act as anodes, suspended in a warm solution of dilute sulfuric acid. Thin sheets of pure copper are interleaved in the thick slabs. The latter act as cathodes. As a result of electrolytic action, copper from the anodes is deposited at the cathodes and impurities are deposited at the bottom of the bath.

Forms or Grades

Cathode Copper > 99.9% pure copper which is produced by electrolytic refining and used as a casting material and as a raw material for the production of high conductivity copper alloys.

Electrolytic Tough Pitch High Conductivity Copper Produced from cathode copper which is melted and cast into billets and other shapes suitable for working; the oxygen content or 'pitch' is carefully controlled to reduce some effects of impurities.

Fire Refined Tough Pitch High Conductivity Copper Fire refining cannot remove some of the fine impurities that electrolytic refining can accomplish. Conductivity of this grade is therefore slightly less than the latter type. The fire refined type of copper is used widely in electrical conductors. Tough pitch coppers are drawn into wire and tubes and used in circuit breakers and electrical power systems.

Oxygen-free High Conductivity Copper Cathode copper which has been further refined by casting wherein oxygen absorption is prevented. This grade of copper is most suitable for flame welding and brazing. It is also suitable for impact extrusion and has good electrical and thermal conductivity. Utilized in making glass-to-metal seals.

Arsenical Copper With the addition of about 0.5% arsenic, the tensile strength of copper is improved and it may be used up to 300 °C without appreciable loss of strength. Corrosion resistance is improved but electrical and thermal conductivity are greatly reduced. Applications for this type of copper are mainly in heat exchanger units such as boiler tubes and radiators.

Copper Base Alloys

Copper base alloys are classified into two major types:

    Type 1 — those which contain small amounts of other metals;
    Type 2 — those which contain large amounts of other metals.

Type 1 alloys include silver copper, cadmium copper, chromium copper, tellurium copper, beryllium copper, and copper – nickel – silicon combined. Type 2 alloys include brasses and bronzes.

Type 1 Alloys

Silver Copper   With the addition of about 0.08% silver, the softening temperature of copper can be increased from about 200 °C to 350 °C, thus the hardness and strength will not be reduced by soldering or low temperature heating. The conductivity change is negligible and creep resistance is increased. Typical applications of silver copper include commutator segments and radiator parts.

Cadmium Copper   When 1% cadmium is added to copper, the softening temperature is raised and strength, toughness, and fatigue resistance are improved. Applications of cadmium copper include long-span overhead conductors, contact wires, and resistance welding electrodes. Cadmium copper is used for electrical wiring in aircraft because of its flexibility and resistance to vibration damage. The addition of small amounts of cadmium has only a slight detrimental effect on conductivity.

Chromium Copper   The addition of about 0.5% Cr has little effect on conductivity but improves strength and hardness and results in a heat treatable alloy.

Tellurium Copper   About 0.5% Te is added to copper to form a readily machinable and high conductivity alloy. Tellurium is insoluble in copper and is dispersed uniformly in the molten alloy. It remains in fine particles in the solid alloy, so that it breaks up chips during machining and makes cutting easier.

Beryllium Copper   Beryllium is alloyed with copper when strength is more important than conductivity. Two beryllium – copper alloys are of industrial importance: the stronger and harder alloy contains 2% Be; but the alloy containing 0.4% Be and 2.6% Co is less expensive. With beryllium contents up to about 3%, Be – Cu alloys are precipitation hardenable. Beryllium – copper is used for corrugated diaphragms, flexible bellows, and for bourdon tubes. It is also used for nonsparking cold chisels and hacksaw blades.

Copper – Nickel – Silicon Alloys   Nickel and silicon alloyed in copper form nickel silicide and result in precipitation hardenable alloys. The total nickel and silicon content depends upon the alloy application, but is usually between 1 and 3%. These alloys have good thermal and electrical conductivities, and good resistance to scaling and to oxidation at high temperatures; they also retain their mechanical properties at fairly high temperatures.

Type 2 Alloys

Brasses are alloys of copper with up to 50% zinc; they may also contain small

27

quantities of tin, manganese, lead, nickel, aluminum, and silicon. As the zinc content is increased, strength, hardness, and ductility increase while thermal and electrical conductivities decrease.

There are three types of brass:

1. *Low*: 5 – 20% Zn, extremely ductile, malleable, and easily worked alloys at room temperatures.
2. *High*: 20 – 45% Zn; alloys which can only be lightly cold-worked.
3. *Alloy*: alloys which contain up to 4% lead to improve machinability at the expense of reduced ductility and strength.

True bronzes are alloys of copper and tin with small quantities of other elements such as nickel and lead. True or tin bronzes are used with < 8% Sn in cold workable sheet and wire forms and in coinage; with 8 – 12% Sn in gears, bearings, and marine fittings; with 12 – 20% Sn, mostly in bearing applications; and with 20 – 25% Sn in cast bells. The main types of other bronzes are as follows:

Phosphor Bronze   These have tin contents ranging from 1.25 to 10%. They have excellent cold-working characteristics, high strength, hardness, and endurance properties, low coefficients of friction and excellent corrosion resistance, and hence find wide use as springs, diaphragms, bearing plates, bellows, and fasteners.

Aluminum Bronze   Contain less than 12% aluminum. All have high strength and wear resistance and possess excellent corrosion resistance, particularly to acids.

Cadmium Bronze   Alloys of copper and cadmium with or without small additions of tin are used primarily for electrical conductors where high conductivity and improved strength and wear resistance over that of copper are required.

Silicon Bronze   The silicon bronzes are an extremely versatile series of alloys having high strength, exceptional corrosion resistance, excellent weldability, and excellent hot and cold workability.

Nickel Bronze   Small amounts of nickel are often added to tin bronzes to give better mechanical properties to castings. They resist wear and corrosion and retain strength at elevated temperatures. Uses include valves and pump parts.

Columbium

Columbium is also called niobium. It occurs naturally with tantalum (another refractory metal) in the minerals columbite and tantalite. The major sources of columbium are in the Belgian Congo, Australia, Nigeria, Brazil, Malaya, Zimbabwe-Rhodesia, and North America.

Initial interest in columbium was stimulated by its low neutron absorption cross-section and high temperature strength. On the basis of strength/weight ratio, columbium alloys are superior to nickel base alloys and cobalt base alloys at temperatures above 900 °C and comparable to molybdenum base alloys to 1400 °C. Characteristics of columbium include:

high melting point ($\sim$ 2500 °C)
moderate density
excellent ease of fabrication
oxidation resistance
low modulus of elasticity
susceptibility to embrittlement by carbon, oxygen
    and nitrogen

Alloying improves the mechanical properties of columbium greatly, but at the expense of a small decrease in fabricability. High temperature strength and oxidation resistance are also increased by alloying. The latter properties of columbium alloys are superior to those of other refractory metals and alloys.

Unlike many other refractory metals, columbium alloys do not form liquid or volatile oxides below 1400 °C, and the scale formed is relatively adherent. Columbium alloys are serviceable for limited periods of time at temperatures of about 1200 °C, but protective coatings are required for longer times or for higher temperatures.

The wet corrosion behavior of columbium at room temperature is excellent and its resistance to liquid metals favors its use in nuclear reactors. Unalloyed columbium is resistant to the following metals to the indicated temperatures: gallium (400 °C), bismuth (560 °C), mercury (600 °C), lead (980 °C), lithium (1000 °C), and sodium – potassium alloys (1000 °C).

Applications

Columbium alloys are employed extensively in components or structures for nuclear reactors, gas turbines, ramjets, airframes, and missiles. At elevated temperatures the metal also absorbs gases, hence its use in high-vacuum tubes.

Columbium is used in stainless steels in small amounts to inhibit intergranular corrosion. In chromium steels it reduces air hardening, increases impact strength, reduces creep, shortens annealing time, and improves oxidation resistance.

Columbium alloys are used in high temperature parts in turbines and missiles. Columbium – tungsten alloys (11% W) are used in instruments and possess good elevated temperature tensile strength.

Columbium – zirconium alloys known as Fansteels are used for missile, aircraft, space vehicle, and nuclear parts.

Columbium selenide is an adhesive lubricating film which is more electrically conductive than graphite.

Fabrication

Pure columbium is considered to be one of the most workable of the refrac-

29

tory metals and it may readily be forged, rolled, swaged, drawn, and stamped. In the primary or mill fabrication, an ingot is hot worked by forging or extruding, following which the surface is conditioned to remove the contaminated layer, annealed in vacuum or an inert atmosphere, and then cold worked (with intermediate anneals, if required) to final shape and size. Columbium containing less than a total of 0.12% combined oxygen, nitrogen, and carbon can be cold worked to a reduction of over 90% in cross-sectional area.

Secondary fabrication is performed cold to avoid oxygen contamination, and lubricants are used to minimize galling or seizing on the working tools. Straightforward procedures are used for making tubes, cups, flanged or flared sleeves, cones, and other forms. Machining performance is most satisfactory with high speed tools, rapid rates, and a strong flow of water-soluble oil coolant.

Columbium alloys are more difficult to work than the unalloyed metal. The technology for joining columbium alloys is still being developed. Tungsten inert-gas welding is the most widely used process, but work is also being carried out on other techniques.

Lead

Lead is one of the most stable metals. The processes used in its production are essentially the same as those used for zinc. Lead sulfide ore is converted to lead oxide which is reduced to metallic lead by a continuous process in a blast furnace. Refining of blast furnace bullion yields 99.99% pure lead. Lead has an extremely high density with outstanding corrosion resistance and a low melting point and is soft and workable. It is, however, a poison.

Small amounts of silver, copper, or tin, when added to lead, improve creep and fatigue resistance and tensile strength. The addition of less than 0.1% tellurium enables lead to be work hardened. Small additions of alkali metals are used to harden lead and to increase its mechanical properties at elevated temperatures. Addition of 8 – 12% antimony improves hardness to the extent that machining of screw threads is possible.

The greatest use for lead in Australia is in the manufacture of lead storage batteries for a wide range of duties. Next in volume is the use of lead in the sheathing of underground cables. Large quantities of tetraethyl and tetramethyl lead are also used as anti-knock additives to gasoline. The most commonly used metal solders are lead – tin alloys. Other alloying elements such as silver, bismuth, and antimony can be used for special purposes. By adjustment of the lead alloy composition, it is possible to produce solders with a wide range of melting points and physical properties. Lead's properties of density and limpness make it an ideal sound barrier. These same properties make it an excellent absorber of nuclear radiation. Red lead priming paints for iron and steel are used extensively. The pliability and corrosion resistance of lead make it an ideal material for collapsible tubes such as toothpaste tubes. Where foodstuffs are contained, lead tubes are lined with tin. 'Type metals' used in printing are alloys of lead, tin, and antimony, which allow high speed, low temperature castings of printer's type.

30

Magnesium

Magnesium is produced by electrolysis from sea water, dolomite, or brines. It is usually extracted from sea water, which contains about 0.13% Mg in unlimited supply. Lime is added in the form of roasted sea shells treated chemically to precipitate magnesium hydroxide. The hydroxide is filtered and treated with hydrochloric acid, resulting in magnesium chloride, which is dried and placed in an electrolytic cell. Magnesium results and is then formed into ingots.

Magnesium is a silvery white metal and is readily cast. It is the world's lightest structural metal, being two-thirds the density of aluminum. Because of its hexagonal close-packed lattice structure, magnesium is difficult to shape at room temperature.

Magnesium has good surface stability under ordinary atmospheric conditions. It resists alkalis, oils, several acids, solvents, and most organic chemicals. It is nongalling and nonmagnetic; snow, ice, sand, and other granular substances do not adhere to it. Pure magnesium, like many metals, must be alloyed with other elements to provide strength. Magnesium alloys have high strength-to-weight ratios. The alloys are easy to machine and are readily fabricated by most metal working processes such as welding. The alloys have excellent formability at elevated temperatures. Magnesium is usually alloyed with aluminum (6 – 12%), zinc (0 – 3%), and manganese (0 – 2%), and with small additions of silicon, cadmium, tin, lithium, zirconium, and cerium. Magnesium alloys possess good static and fatigue strengths and ductility qualities, good dent resistance, high damping capacity and toughness, and low moduli of elasticity.

Magnesium alloys are dimensionally stable at 95 °C. Some cast Mg – Al – Zn alloys may undergo permanent growth if used above 95 °C for long periods. Mg – rare earth and Mg – thorium alloys are dimensionally stable to above 320 °C. Other properties of magnesium alloys include good conductivity, low specific heat, and good thermal diffusivity.

Magnesium alloys and tempers are designated according to ASTM (American Society for Testing and Materials) specifications. A typical designation would be alloy number AZ61A-F, where:

1. A indicates that principal alloy is aluminum
2. Z indicates that the second most prevalent alloying element is zinc
3. the 6 indicates that 6% Al is used
4. the 1 indicates a range in zinc content
5. A indicates the first alloy developed having this composition
6. the F indicates an 'as-fabricated' condition

Uses

Magnesium is used extensively in satellites and space probes. New alloys, protective treatments, and fabrication processes are still being developed. Other uses include:

airborne equipment

automobile wheels
engine parts such as gear boxes
electronic and X-ray equipment
equipment and machine housings
materials handling equipment
missile and aircraft structure
portable power tools

Molybdenum

The element molybdenum is a member of the refractory metals. It is obtained from its chief ore molybdenite, MoS_2. The metal is a soft, opaque, dark gray, graphite-like material. The USA has been the chief producer since 1952.

Production and Processing

The refining process of molybdenum involves roasting the ore and converting the oxide obtained to ammonium molybdate, which is purified by crystallization. Ignition of salt gives an oxide of high purity. This is reduced by hydrogen to a metal powder which is then pressed, swaged, and drawn into wire form. Molybdenum may also be recovered economically as a by-product of copper ore mining.

Consumable arc-melted molybdenum is finer-grained and frequently more ductile at room temperature than the purer electron beam-melted metal. Molybdenum powder may be slip cast to desired shapes and sintered to obtain shapes such as a crucible. The powder may be directly rolled to flat sheet, sintered, and hot rolled to various gages, even to foil.

Molybdenum is ductile and softer than tungsten and is readily drawn into very fine wire. It cannot be hardened by heat treatment but can be hardened by cold working. The rolled metal has a tensile strength comparable to high strength steel and possesses high corrosion resistance.

Properties

The density of molybdenum is nearly four times that of aluminum and 90% of that of lead. Its melting point is about 2600 °C. Electrical conductivity is 35% of that of copper, therefore molybdenum is a relatively good conductor. Thermal expansion is low and heat conductivity is high, being twice that of iron.

The main disadvantage of molybdenum is oxidation at high temperatures. Even though is has a good high temperature strength, it must be protected from oxidation. This presents difficulties, for it must also be protected against reactions with silicon or carbon, depending upon service environment. A practice designed to take advantage of high strength qualities is to use hydrogen atmosphere as protection for molybdenum at high temperatures.

32

Molybdenum can be resistance-welded by spot, projection, seam, and upset-butt welding. Very short welding times are required at high power levels. Welds, because of grain growth tendency, may be slightly brittle and of low strength unless proper controls are employed.

Uses

Uses fall into two classes, the first with the pure metal and the second with alloys.

Pure Molybdenum Because of relatively good electrical properties, molybdenum in nearly pure form is used to make anode cups for electronic tubes, support members in radio and light bulbs, heating elements for electric furnaces, and arc-resistant electrical contacts. Molybdenum windings protected by hydrogen are used for high temperature furnaces and, with the increased use of vacuum techniques, quite complex assemblies are now becoming common.

The excellent bond between sprayed molybdenum coatings and metals such as steel, aluminum, and magnesiun eliminate the need for elaborate surface preparation prior to metallizing. Wide use is made of sprayed molybdenum for good wear resistance and it is used for salvaging worn shaft bearings, crankshafts, and diesel transmission housings. Apart from its maintenance benefits, it is also used for new parts such as piston rings.

Molybdenum is resistant to hydrofluoric acid and to a wide range of molten metals including lead, copper, mercury, sodium, and potassium. Resistance to the two last-mentioned elements is of particular importance in nuclear plants which use them as heat transfer media. Molybdenum is also used in high temperature structural parts of jet engines and missiles.

Alloys The use of molybdenum for alloying began as a scientific curiosity, but has evolved to become one of the basic alloying elements for high strength, low alloy steels in addition to its growing use in nonferrous alloys. Titanium added to molybdenum in amounts as small as 0.5% increases tensile strength significantly. The presence of molybdenum in steel increases the strength, toughness, and wear resistance. Molybdenum is also the most important alloying element for minimizing temper brittleness and thus assuring attainment of maximum toughness in steels. As an alloying element in stainless steels, molybdenum improves corrosion resistance. In tool steels it improves hardness and minimizes embrittlement caused by ageing.

Nickel

Extensive nickel ore mining is carried out in New Caledonia and in Ontario, Canada. The main steps in the extraction process are as follows:

1. Sulfide ore is crushed and sulfides are separated by flotation.
2. Nickel and copper sulfides are further separated by another float.

3. Nickel sulfides are roasted, which removes some sulfur, and then undergo a Bessemer type process. Impurities are removed by oxidation.
4. A mixture of molten copper, nickel, and iron sulfide is obtained after further chemical treatment.
5. Segregation occurs during cooling. Unwanted products are removed mechanically. The nickel sulfide is then sintered, forming an oxide, and pure nickel is formed by means of a reduction process using hydrogen and electrolysis.

Properties

Commercially pure wrought nickel is a grayish-white metal capable of taking a high polish. Due to its combination of attractive mechanical properties, corrosion resistance, and formability, nickel and its alloys are used in a variety of structural applications usually requiring corrosion resistance.

Nickel is used as an undercoating for parts that are to be chromium plated. The pure nickels have > 94.0% nickel content in combination with aluminum, manganese, or silicon.

When combined with cobalt, the alloy is heat resistant and is used in the chemical industry. Alloyed with manganese, nickel becomes resistant to oxidation and reduction at elevated temperatures. When alloyed with aluminum, it develops high strength as well as corrosion resistance. Nickel when used with copper is highly resistant to deterioration by salt water or sulfuric acid. Alloyed with chromium, nickel is resistant to high temperature, oxidation, and corrosion. When alloyed with titanium or aluminum, nickel will age harden and is used in electroplating.

Classification

The *Monel* metals are essentially combinations of nickel and copper. They possess high corrosion resistance and strength. They are, therefore, used to resist corrosion due to sulfuric acid, caustic soda, and salt water.

The *Inconel* metals are nickel alloys that contain substantial chromium and iron. They have good strength and have high resistance to corrosion by chemicals at high temperatures.

The *Hastelloy* metals are nickel alloys that contain various combinations of chromium, molybdenum, and iron. They have excellent resistance to oxidizing chemicals, even at elevated temperatures.

The *Illium* nickels are nickel alloys with the iron component of the Hastelloy metals being replaced by copper. They have good corrosion resistance and strength.

Many of the above nickel alloys are available in a wide range of forms and can be hot or cold worked. Other processes include annealing, drawing, forging, spinning, and joining by soldering, brazing, gas welding, and resistance welding. Machining of most nickel alloys requires high speed tools and flooding with cutting compounds.

Uses

The principal uses of nickel alloys are:

1. for corrosion service
2. in cryogenics
3. in high strength and elevated temperature applications

Numerous applications therefore exist in drums for shipping chemicals, food processing equipment, gas turbine parts, plating for decorative purpose (i.e., jewellery, plumbing fixtures, automobile trims and bumpers), manufacture of dies for plastics requiring good corrosion resistance and high finish, and use in transducers, electrical and electronic components, and water storage tanks.

Recent Developments

Extensive testing was undertaken for material selections in a wide range of installations and equipment used to scrub effluent gases in chemical and electrical power plants. The materials selected had to withstand high temperatures, corrosion, and erosion and be strong enough to resist high pressure or vacuum. Effective materials found were the medium grade Hastelloy 'G' and Incoloy 825 alloys and the high grade Hastelloy 'C' and Inconel 625 alloys.

It was found recently that a polyester-reinforced glass fiber sonar dome could be made resistant to marine fouling attachment by the addition of a $90-10$ Cu $-$ Ni flake dispersed in the polyester resin without degrading the sonar's receiving sensitivity.

Gold $-$ nickel alloys are used for brazing, although they are limited to alloys near the 82.5% Au $-$ 17.5% Ni composition, where the liquidus and solidus converge to give a fixed melting point. They have been used extensively in the aerospace industry for brazing turbine blades, fuel pipes, etc.

References

STREET, A.C, and ALEXANDER, W. *Metals in the Service of Man*, 5th edn, Penguin Books, Harmondworth, England and Baltimore, 1972

HUME-ROTHERY, W., SMALLMAN, R.E., and HAWORTH, C.W., *The Structure of Metals and Alloys*, 5th edn, Institute of Metals, London, 1969

KEMPSTER, M.H.A., *Materials for Engineering*, Hodder and Stoughton, London, 1964

Materials Performance, Volume 14, Number 4, April 1975, National Association of Corrosion Engineers, Houston, Texas

Metals Australia, Volume 8, Number 4, May 1976 Australian Institute of Metals

POLLACK, H.W., *Materials Science and Metallurgy*, Reston Publishing, Virginia, 1973

Precious Metals

The precious metals include platinum, palladium, iridium, osmium, rhodium, ruthenium, gold, and silver.

Platinum

This whitiesh-gray metal is more ductile than silver, gold, or copper and is heavier than gold. It occurs in small flat grains, usually in alluvial sands. The native metal generally contains other metals of the platinum group. The chief sources are Russia and Columbia, with small amounts from Alaska, Canada, and South Africa, where it is obtained from copper – nickel ores.

Platinum is very ductile and malleable. It is resistant to acids and alkalis but dissolves in aqua regia. The metal is widely used in jewellery, but because of its heat and chemical resistance it is also valued for electrical contacts, resistance wire, thermocouples, standard weights, and laboratory dishes. It is generally too soft for use alone and is almost always alloyed with harder metals of the same group such as osmium, rhodium, and iridium.

Because of its high resistance to atmospheric corrosion even in sulfur environments, platinum in the form of coatings is used on springs and other functioning parts of instruments and electronic devices where precise operation is essential. Platinum – rhodium alloys containing 10% rhodium are used in thermocouples for temperatures above 1100 °C.

Palladium

Palladium is a rare metal found in the ores of platinum. It resembles platinum but is less dense and has a silvery luster. Even though it is only half as plentiful, the metal is less costly than platinum. It is highly resistant to corrosion and to attack by acids but, like platinum, is dissolved in aqua regia. The metal is ductile and is readily hot or cold worked to very fine sections.

Palladium alloys readily with gold to form white golds as well as alloying in all proportions with platinum to form a harder material. Although palladium has low electrical conductivity, 10% that of copper, it is valued for its resistance to oxidation and corrosion. Palladium-rich alloys are widely used for low voltage electrical contacts and in printed circuit boards.

A few alloys are: palladium – silver alloys (30 – 50% silver), used for relay contacts; palladium – silver alloys (25% silver), used as a catalyst in powder or wire mesh form; and palladium – copper alloys (40% copper), used for sliding contacts. Palladium alloys are also used for instrument parts and wires, dental plates, and fountain-pen nibs. They are valued for electroplating as they have a fine white tarnish resistant color.

Palladium powder is made by chemical reduction and has a purity of 99.9%. This powder goes into solid solution with hydrogen, forming palladium sponge, which is used for some gas lighters.

Iridium

Iridium is a grayish-white metal of extreme hardness. It is insoluble in all acids. The metal is found in the nickel – copper ores of Canada, pyroxinite deposits of South Africa, and platinum ores of Russia and Alaska. It occurs naturally with the metal osmium as an alloy known as osmiridium (30 – 60% osmium) which is used for making fountain-pen points and instrument pivots. Iridium is employed as a hardener for platinum, the jewellery alloys usually containing 10%. Iridium plating is used on molybdenum to protect against oxidation at high temperatures.

Above 600 °C iridium tarnishes and above 1000 °C it forms a volatile oxide. Iridium wire is used in spark plugs because it resists attack of leaded aviation fuels. Iridium – tungsten alloys are used for springs operating at temperatures to 800 °C.

Osmium

Like iridium, osmium has a density twice that of lead. It also has a very high modulus of elasticity (550 GPa). It is the heaviest and hardest of all precious metals (Vickers 1000, cold worked). The metal forms solid-solution alloys with platinum, having more than double the hardening power of iridium and platinum. The metal is very brittle and a high melting point making fusion and casting difficult. Osmium is not affected by common acids and is used in pen-tips and instrument pivots. The tetroxide gas formed from osmium is highly poisonous.

Rhodium

Rhodium is found in platinum ores in Canada and South Africa. It is insoluble in most acids but is attacked by chlorine at elevated temperatures and by hot fuming sulfuric acid. Liquid rhodium dissolves oxygen and ingots are made by argon-arc melting. At temperatures above 1200 °C it reacts with oxygen to form rhodium oxide. Rhodium is used to make resistance windings in high temperature furnaces, for high temperature thermocouples, as a catalyst, and for laboratory dishes.

The most important alloys of rhodium are rhodium – platinum. They form solid solutions in any proportions, but alloys of more than 40% rhodium are rare. These alloys are used for thermocouples and in the glass industry, partic-ularly in mirrors, since they are easily worked and do not tarnish or oxidize at high temperatures.

Ruthenium

The hard, silvery-white metal ruthenium is obtained from the residue of platinum ores by heat reduction of ruthenium oxide in hydrogen. Ruthenium is the most chemically resistant of the precious metals and is not dissolved by aqua regia. It is used as a catalyst to combine nitrogen in chemicals.

By adding 5% ruthenium to platinum, hardness is increased significantly and the alloy's electrical resistivity is twice that of pure platinum. The metal is practically unworkable and readily forms a gaseous oxide on heating in air. Ruthenium – platinum alloys (5 – 10% ruthenium) are used for electrical contacts and wires, chemical equipment, and jewellery.

Gold

Gold was discovered about 4000 BC. It is of well-known unique yellow color and is not tarnished by air or by heat. Gold is mined mainly in South Africa, which provides more than 50% of the world's production. Gold occurs with quartz and does not have to be separated from ore to be produced commercially.

The metal is remarkably ductile and malleable, hence it can be easily cold worked. As well as having high thermal and electrical conductivities, it combines chemically only with hot alkali cyanides and free chlorine. Its electrical resistance is influenced by impurities. Nearly pure gold is used as the world's monetary standard.

Gold alloys in all proportions with silver and is used mainly for jewellery. Alloying with copper produces a harder material with a reddish yellow color.

The major use of gold alloys is in the dental industry. A specialty use of gold is that of thin films applied by thermal evaporation in vacuum to the plastic visors of astronauts' helmets to reduce the glare from sunlight while permitting good visibility. Television cameras and batteries in space applications have also been covered in gold to maintain even temperature.

Silver

Silver was discovered about 2000 BC. It exhibits similar properties to gold in that it is stable to the effects of atmosphere and heat. Mined in North, Central, and South America, silver is separated from its ores of lead and copper by smelting. Similar to gold, it has high thermal and electrical conductivities; its electrical conductivity exceeds than that of copper. The white soft metal is very ductile and malleable and exhibits a high affinity for oxygen, hence special precautions must be taken when welding. The metal is readily corroded when in contact with sulfur.

Silver alloys with both zinc and copper, an important example being silver solder. Silver is used in certain goods processing as container linings, in the photographic industry, and in heavy duty bearings, where it is alloyed with 3 – 5% lead. In dentistry, silver amalgams are used because they are plastic and workable for a short time and then become very hard. Silver also finds use in silvered microcapacitors and other components.

References

PATTERSON, W.J. *Materials in Industry*, Prentice-Hall, Englewood Cliffs, N.J., 1968

WILKINSON, W.D. *Properties of Refractory Metals*, Gordon and Breach, New York, 1969

Metals Australia, Volume 7, number 9, October, 1975; Volume 8, Number 1, February, 1976; Volume 8, Number 2, March, 1976; Volume 8, Number 3, April, 1976; Volume 8, Number 4, May, 1976; Australian Institute of Metals

Rare Earth Metals

The 15 metals occurring in the periodic table from lanthanum to lutetium, plus scandium and yttrium, which are not in this series but occur together in nature, are together called the 'rare earths'. One-sixth of the known elements are rare earths and they comprise one-quarter of all metals. (See Table 8.)

Table 8   The Rare Earths

Name	Symbol	Atomic number
Scandium	Sc	21
Yttrium	Y	39
Lanthanum	La	57
Cerium	Ce	58
Praseodymium	Pr	59
Neodymium	Nd	60
Promethium	Pm	61
Samarium	Sm	62
Europium	Eu	63
Gadolinium	Gd	64
Terbium	Tb	65
Dysprosium	Dy	66
Holmium	Ho	67
Erbium	Er	68
Thulium	Tm	69
Ytterbium	Yb	70
Lutetium	Lu	71

This series of metals is frequently divided into two groups based on their atomic numbers and chemical properties:

1. 'Light rare earths', of atomic number 57 – 63
2. 'Heavy rare earths', of atomic number 64 – 71, 21, and 39

Occurrence

The word 'rare' as used in rare earth elements does not imply that the metals are extremely hard to find due to their scarcity, but that they are difficult to extract. Their ores are mostly carbonates. Thulium, the least abundant of these metals, is more common than cadmium, mercury, or even silver. Of the lanthanides, lanthanum, cerium, praseodymium, and neodymium make up 90% of the naturally occurring minerals, while the heavier elements make up the remainder of these minerals.

Commercial rare earth metals contain 94 – 99% rare earth metal with traces of calcium, carbon, aluminum, and up to 1% silicon and 1 – 2% iron. Scandinavia, India, the Soviet Union, and the USA are the best known deposits. Rare earths, however, are found in many other places.

Separation and Form

Prior to the advent of ion exchange separation techniques, rare earth metals were expensive. The ion exchange technique is now used and produces the rare earth metals, but only in ingot form.

Mischmetal, the major commercial form of the rare earths, is available at a cost of $8/kg (at the time of writing) in plates weighing 20 – 30 kg each. At a normal purity of 99%, the cost of most of rare earths is $250 – 800/kg. Lutetium, thulium, terbium, and europium are more expensive.

Properties

Rare earths are soft and malleable metals and increase in hardness as atomic number increases. As impurities are added, their hardness and malleability change.

All the metals are paramagnetic and, in general, are uniquely good thermal conductors but poor electrical conductors. The melting points (with the exception of cerium, europium, and ytterbium) increase with increasing atomic number, with the melting point of lutetium being about twice that of lanthanum.

For commercially pure rare earth metals (99.5% pure), yield strengths are in the range 100 – 350 MPa and elastic moduli are 40 – 80 GPa. Strengths at 400 °C are about half those at room temperature. Yttrium provides an interesting combination of properties: density similar to titanium, a melting point of 1548 °C, and it forms one of the most stable hydrides.

All rare earths can be hot worked, while some can be cold worked. Lanthanum, cerium, and europium oxidize rapidly under atmospheric conditions while others form protective oxide films.

Applications

Mischmetal, a combination of rare earth metals, was originally used for pyrophoric alloys (i.e., they burn spontaneously in air) but is now used in cast iron because it opposes graphitization and produces a more malleable iron. It is used in steel to remove sulfur and oxides and degasifies it. In stainless steels, it is used as a hardening agent. As alloying elements, rare earths increase the life of heating elements containing nickel and chromium. They also increase the workability of steels and increase the oxidation resistance of some ferrous and nonferrous alloys.

Oxides of the rare earths are used in glass coloring or decoloring and in glass polishing. Certain rare earths and their oxides have uses in high temperature applications and as neutron absorbers. A radioisotope of samarium has been used for gammagraphing engine castings. Other applications of rare earths include chemical catalysts, permanent magnets, and semiconductors.

Refractory Metals

The refractory metals are chromium, columbium, molybdenum, tantalum, tungsten, and vanadium. Columbium and molybdenum have been discussed earlier in this chapter. Each of the remaining metals is treated under properties and applications.

Chromium

Chromium is silvery-white with a blueish tinge and is used for oxidation and corrosion resistance in stainless steels, heat-resistant alloys, high strength alloy steels, and for wear-resistant electroplating. Its compounds are used in pigments, chemicals, and refractories. The melting point of chromium, 1550 °C, is the lowest of all the refractory metals; it is an extremely hard metal. It is resistant to oxidation and inert to nitric acid but dissolves in hydrochloric acid. At temperatures above 820 °C it is subject to an intergranular corrosion known as 'green rot'.

Chromium occurs in nature only in combination. Its chief ore is chromite, from which it is obtained by reduction and electrolysis. Chromium lacks ductility and is susceptible to nitrogen embrittlement. Alone it is not used generally as a structural metal.

Tantalum

Tantalum is a white lustrous metal resembling platinum. It is highly acid-resistant and is classed as a noble metal. Its melting point is almost 3000 °C. Tantalum is very ductile and can be rolled without annealing.

At very high temperatures tantalum absorbs 740 times its volume of hydrogen, producing a coarse, brittle substance. Tantalum can be hardened to about 600 Brinell by heating in air. It will hold a fine cutting edge on tools. By adding silicon to tantalum, its hardness can be made close to that of diamond. Other uses of tantalum include electric light bulb filaments, radio tubes operating at high temperatures, surgical instruments, and current rectifiers.

Tantalum – tungsten alloys are used in rocket motor parts and have melting points to 3500 °C. T-111 (8% tungsten) has high corrosion resistance and high tensile strength. T-222 (9.6% tungsten) has a greater tensile strength than T-111 and is used in elements to heat acid baths.

Tungsten

Tungsten is a heavy white metal and possesses the highest melting point of all metals, 3410 °C. It is widely distributed in small quantities in nature, being about half as abundant as copper. The metal is usually obtained as a powder by reduction of the oxide. Tungsten resists oxidation at very high temperatures and it is not attacked by nitric, hydrofluoric, or sulfuric acid solutions. The metal is brittle and difficult to fabricate.

Tungsten has a wide usage in alloy steels, magnets, heavy metals, electric contacts, light bulb filaments, rocket nozzles, and electronic applications. Parts, rods, and sheet are made by powder metallurgy using tungsten powder of 99.9% purity, and rolling and forging are performed at high temperatures. The rolled metal and drawn wire have exceptionally high tensile strength and hardness.

Tungsten wire for spark plug and electronic use is made by powder metallurgy. Tungsten whiskers are used in copper alloys to provide strength. Cobalt – tungsten alloy (50% tungsten) is used in plates that retain high hardness at elevated temperature.

Vanadium

Vanadium is a pale gray metal with a silvery luster which is found in Peru, Zimbabwe-Rhodesia, South West Africa, and the United States. Vanadium melts at 1000 °C. It does not oxidize in air and is not attacked by hydrochloric or dilute sulfuric acid. It is marketed usually in 99.5% pure cast and machined ingots and buttons. Vanadium is expensive. Its greatest use is in alloying, its function being to enhance properties developed by other alloying elements. Vanadium is used in making high speed tool steel and high strength and outstanding magnetic materials such as Perminvar (Fe – Co – V). Vanadium salts are used to color pottery and glasses.

Tin

The chief source of tin is cassiterite, with Nigerian columbite being a lesser source containing 6% tin oxide. The major tin producing countries are Indonesia, Malaya, Bolivia, China, and Nigeria.

The physical appearance of tin is a silvery-white lustrous metal, with a blueish tinge. It also has noticeable characteristics of softness and malleability. It is noteworthy that about 40% of the world's consumption of tin is as a coating material for steel 'tin' cans.

Tin is comparatively easy to reduce to metal from the oxide by pyrometallurgy. The washed and dried ore is heated with powdered coal and lime to form a slag in a reverbatory furnace at 1200 °C. In this process, the tin tends to be absorbed into the slag and thus the process needs to be repeated. The impure tin is then refined by melting in a sloping hearth which allows the metal to run off. Ninety-nine percent pure tin is obtained when zinc present in oxidized. Electrolytic refining produces a tin of greater purity.

42

Properties

Tin can be rolled into very thin foils. It melts at 230 °C and has a low tensile strength. Tin is slightly harder than lead. Its electrical conductivity is about one-seventh that of silver. Tin is resistant to atmospheric corrosion and dissolves in mineral acid. Below 20 °C 'tin pest' occurs, which is the transformation of the metal to a gray powder, thus low temperature applications are to be avoided.

The corrosion resistant properties of tin are due largely to 'Protectalin', which is the name given to the thin invisible oxide film on tin plate that protects it from sulfur staining.

'Block tin' is virgin tin cast in molds. Very small traces of other metals change the physical properties of tin drastically. Lead softens the metal and arsenic and zinc harden it. The addition of 0.3% nickel doubles the tensile strength, 2% copper increases the strength by 150%. Small amounts of impurities broaden the melting points.

Tin can be cut readily by a knife, but the process is hindered by the tin sticking to cutting edges. Hammering the metal has the effect of hardening it. Tin is classified according to its purity:

highest purity	$> 99.9\%$
refined	$99.75 - 99.9\%$
common	$99.0 - 99.75\%$

Cost of the metal is about four times that of aluminum.

Alloys

Alloys account for about half of the world's consumption of new tin. These cover a wide composition range and many applications because tin alloys readily with nearly all metals.

Tin-plate is soft steel plate having a thin coat of highest purity tin on both sides. It is made by hot dipping. Classifications are as follows:

Coke tin-plate	Minimum amount of tin to protect and brighten.
Charcoal plate	Heavier costs of tin in various thicknesses.
Terne plate	Coatings on steel plate with alloys of lead and tin. Used in automobile fuel tanks. Compositions vary between $12 - 88\%$ Sn − Pb and $50 - 50\%$ Sn − Pb.

Solders are primarily of lead − tin or lead − tin − antimony alloys with up to 94% Sn. Bronzes are combinations of copper and tin with up to 10% tin being commonly used.

Brasses are basically copper − zinc alloys, although some brasses may contain small amounts of aluminum or tin in addition to Cu and Zn.

Babbitts are tins containing $4 - \$\%$ each of copper and antimony to give compressive strength and a structure necessary for a good bearing material.

Pewter is an easily formed tin-based alloy with about 7% antimony, 2% copper, and 10% lead. It has hardness and luster retention, is readily cast, and is used in craft work and jewellery.

All tin alloys are very workable and thus tools for their machining or cutting need not be of outstanding hardness. Most alloys are readily bonded by melting. Eutectic tin – lead solders or low melting fusible alloys are sometimes used on massive parts.

Uses

Less common than tin coatings for cans, but equally important, are uses of tin-plate in toys, food graters, and containers. The main use of electroplated pure tin coatings is as a preparation for easy soldering for printed circuit boards for radios and televisions. Other examples are tinned copper wire coatings used in electrical industries.

Tin can be co-deposited with several other metals including antimony, bismuth, cobalt, copper, gold, iron, lead, nickel, and zinc. There is no special difficulty in many cases of co-deposition of more than one metal with tin. Despite the large number of alloy platings which can be made with tin, only a few have attained commercial acceptance as plated finishes. The important alloys are tin combined with copper, lead, nickel, or zinc.

The most widely used of all tin alloys are solders in which tin is mixed with varying proportions of lead. The adhesive property of solders on metal surfaces is due to the alloying action of tin, but lead fulfills the useful function of diluting the tin, lowering the melting point, and 'thickening' the molten solder to form a pasty material that can be applied readily.

Bearings for fast-moving machinery of all types—including internal combustion engines, electrical generators and motors, and steam engines—use substantial quantities of babbitts. As well as being used for immersion tinning of metals and sensitizing glass and plastics before metallizing, tin also has application as an opacifier in ceramic enamels to provide an abrasive coating.

The properties of cast iron are improved by the addition of small amounts (0.1%) of tin, in producing uniform high hardness and improved casting characteristics.

A range of titanium – tin alloys play their part alongside other titanium alloys in meeting certain aerospace requirements. A titanium – tin alloy was used in the Apollo moon flights in one of the lunar module cryogenic tanks.

Titanium

Titanium is the fourth most abundant metal on earth, occurring in the ores rutile (TiO_2) and ilmenite ($Fe_2.3TiO_2$). Rutile is a constituent in granite, gneiss, limestone, dolomite, and some beach sands. Ilmenite is found in Australia, Canada, and the USA. The properties of titanium show many similarities to those of silicon and zirconium, while its aqueous solution chemistry shows some resemblance to that of vanadium and chromium.

Properties

Titanium is a silver-gray, paramagnetic metal having strength equal or

44

superior to that of steel, although its density is only about 56% that of alloy steel. It retains its properties in the temperature range 160 – 550 °C. The modulus of titanium is slightly more than half that of steel, almost twice that of aluminum, and three times that of magnesium.

The unique combination of titanium's properties was seized upon by designers seeking to lighten the weight of jet engines and airframes when wrought titanium shapes became available in 1950. Since then titanium has achieved an industrial status that took other metals such as lead, copper, and zinc 40 – 80 years to reach. Of the newer metals, aluminum and magnesium spanned 25 – 30 years before arriving at the production capabilities offered by the titanium industry in 1958. In many respects titanium seems to fill the gap between stainless steel and aluminum.

One of titanium's most outstanding properties is that it can withstand attack by a wide variety of aggressive media. It also has a high resistance to pitting and stress corrosion cracking. Its resistance to corrosion is superior to that of some stainless steels. Titanium, similarly to aluminum, forms an oxide coat which protects it from further oxidation. It is very reactive and must be melted in vacuum. Hot working is also difficult and fusion welding is carried out in an inert atmosphere. Furthermore, titanium has a tendency to work harden, so that cold forming must be carried out in stages with periodic annealing.

Current research on titanium includes means of bonding without creating inhomogeneity in joint zones. Diffusion bonding is a technique that is used for fabrication of complex parts.

Alloys

Titanium alloys are made by conventional steel making methods. At ordinary temperatures they have high strength-to-weight ratios. The main alloying elements added to titanium are aluminum, iron, manganese, vanadium, molybdenum, and chromium.

As regards machinability, titanium alloys rank among the more difficult materials—between tool steels and nickel- and cobalt-based superalloys. The practical difficulties of machining can be solved by powerful, rigid machine tools with slow speeds, heavy feeds, and copious liquid coolant. Difficulties of drilling can be avoided by spark machining.

The susceptibility of titanium alloys to cracking on grinding is very slight, and with suitable non-clogging wheels, grinding and slitting can be performed wet or dry according to circumstances, with account taken of the pyrophoric character of its fine dust.

Uses

Most applications of titanium reflect the material's high strength-to-weight ratio: it is used in aircraft components such as flap tracks, engine mountings, and other parts carrying heavy loads. In gas turbine engines, titanium alloys are used extensively for components such as compressor blades, discs, and casings.

Titanium alloys are competitive substitutes for steel in hydraulic cylinders and pipework and in certain landing gear components and helicopter rotor hubs. In the latter application the material's low modulus—half that of steel—and good fatigue properties combined to enable a hub design in which the hub's elasticity absorbs the blade flapping motion without need for a complex mechanism.

Resistance to corrosion in a wide range of natural or synthetic environments makes titanium an ideal material for many chemical plant units. It is used as a lining or as a primary constructional material for vats, driers, and reactor vessels, for tubular or plate-type heat exchangers, for condensers in power plants, and for pumps, valves, and many other components.

In electrochemical processes, titanium is ideal for making anodizing jigs since the anodic film characteristics permit continued use of the jigs without stripping. Further uses in metal finishing plants include coils and immersion heaters, anode baskets, and jigs. Another potentially large application is in copper refining, where cathode starter plates employ titanium's oxide film as a parting agent as well as a protection against attack by copper sulfate electrolyte.

For general engineering applications, titanium's high strength-to-weight ratio leads to reduction of centrifugal or inertial stresses where it is used in rapidly moving parts such as centrifuges, connecting rods, or steam turbine blades and lacing wire. Chemical and mechanical properties make it attractive also for springs, valves, and seals.

Uranium

Uranium is a very dense nonferrous metal and possesses a silver-gray luster when freshly prepared but readily tarnishes on exposure to air because of oxide film formation. Uranium has at least three isotopes with atomic masses 234, 235, and 238. Isotope 238 makes up 99.3% of the natural metal.

Although uranium is regarded as rare, it is almost as abundant as copper and far more so than zinc, lead, tin, or gold. It does not occur free in nature but is always associated with other minerals. The most common sources are uranite and pitchblende. Canada is presently the chief supplier of uranium. Australia has substantial virtually untapped uranium reserves, estimated to be 13% of the world's total.

Processing

Processing treatments for the production of uranium concentrates and the recovery of uranium compounds are numerous because of the great variety and limits of concentration in its naturally occurring ores. Recovery of uranium requires chemical processing; however, preliminary ore treatment may involve a roasting operation, a physical or chemical concentration, or a combination of treatments. The largest quantities of uranium metal of good quality have been prepared from uranium tetrafluoride by reduction with calcium or magnesium. The recovered metal is quite pure and can be fabricated by conventional methods with due consideration for its chemical reactivity and allotropic transformations.

Uranium can be cast and fabricated into any desired shape by conventional means such as rolling, extrusion, and drawing. The metal is very reactive and oxidizes at moderately high temperatures and must be protected from air during fabrication.

Properties

Uranium has an atomic number of 92. Its melting point is 1130 °C and its boiling point is 3800 °C. Its strength decreases rapidly above 400 °C. Uranium is a poor conductor of electricity, its electrical conductivity being only 2.6% that of copper.

The α phase exists below 662 °C and the metal in this phase is soft and ductile. The β phase exists between 662 and 772 °C. The metal in this case is hard and brittle. A soft γ phase exists above 772 °C. The low strength of this phase is such that small cylinders slump under their own weight.

Uranium has a strong electropositive nature, making it very reactive. In powder form it ignites spontaneously in air. It decomposes water and combines directly with all non-metals except noble gases. It dissolves in hydrochloric acid to leave a black residue of uranium hydroxyhydride. Nitric acid dissolves the metal, but nonoxidizing acids such as sulfuric acid react very slowly.

Uses

The uses of uranium fall into two categories: non-nuclear and nuclear.

Non-nuclear   Due to the long half-life of uranium, it is used to calculate the age of rocks. In the electrical industry it is used as a cathode in photoelectric tubes. Uranium is also used as an alloying element in some steels where it enhances strength and toughness. Uranium alloys (0.5 – 5% U) are used in some high speed tool steels. Uranium is sometimes used as a paint pigment and in glazes for pottery. Small amounts will give a yellowish-green tinge in glass and a reddy-yellow in glazes. It is also used in the medical field, in X-raying devices, and for cancer treatment. Depleted uranium, which is extremely dense (1.9×10^7 g/m³), is sometimes used as counterweights, missile ballast, and shielding.

Nuclear   By far the most important use of uranium is for nuclear power fuel supply. Conversion of 1g of uranium yields energy equivalent to 90 MJ, whereas the same weight of coal releases only 0.03 MJ. As the world's oil supplies decrease, the demand for nuclear fuel and nuclear power stations is increasing. Uranium also has well-known uses in the military field and in nuclear weapons.

Zinc

Zinc is a relatively inexpensive metal which has moderate strength and toughness and outstanding corrosion resistance in many types of service. Its cost as a durable protective coating for steel is lower than that of any organic or inorganic material. The low melting point of zinc makes it especially suitable for diecasting products.

Zinc in the metallic state does not occur in nature. Zinc is estimated to rate 24th in order of abundance of metals. The production of zinc involves roasting zinc sulfide ore to zinc oxide and reducing the oxide to zinc in a coke fired furnace. The zinc is produced as a vapor and condensation yields metal of 98% purity or better. The world's major producer of zinc ore is Canada. The USA, Russia, Australia, and Peru are next in order of importance.

Grading

ASTM gives a specification (B-6) for various grades of zinc according to the amounts of trace elements—Pb, Fe, and Cd—present:

> *special high grade* zinc is used as an alloying element for brass and for as-rolled zinc
> *intermediate form* is also used in a rolled form
> *brass special grade* is used primarily in galvanizing
> *prime western grade* is used for galvanizing, as a leaded brass alloying element, and as an oxide

Applications

Australia uses approximately 100 00 tonnes of zinc per year, which represents an average consumption of about 8.6 kg per person. This is the highest per capita consumption of zinc in the world. More than 60% of the zinc consumed is used in variations of the hot dip galvanizing process to provide coatings for the protection of iron and steel products from corrosion. More than 10% of Australia's zinc consumption goes into the production of brass articles and a similar amount is used in the production of pressure die castings such as carburetor bodies.

Coatings for steel products are also provided by zinc-rich paints or by spraying steel with molten zinc using either powder or wire. Zinc is also used in sheet form for manufacture of dry cell battery cases and for the production of roofing materials, printing plates, and engraving metal. Zinc oxide is an essential ingredient in the manufacture of rubber, plastic, ceramics, and medicinal products and—together with other zinc compounds—is used in the formulation of paints, motor oil additives, soldering fluxes, and other products.

Zirconium

Zirconium is a metallic element and is considered to be a relatively new metal. It is silvery-white in color and has a melting point of 1850 °C. Zirconium is more plentiful than copper and several other common metals. It is obtained principally from two minerals: zircon and baddeleyite. Zircon is found in beach sands in many parts of the world including Queensland. The United States, however, is the main source. Baddeleyite exists in quantity in Brazil only. A characteristic of all zirconium ores is that they contain from 0.7 to 6% hafnium, a metal similar to zirconium and difficult to extract. Hafnium must be removed from zirconium intended for nuclear use, but is seldom objectionable when zirconium is used for other purposes.

Forging, rolling, and most other fabricating processes may be performed on zirconium without difficulty. Ingots are often worked hot in the initial stages. Difficulty may be encountered in drawing zirconium into tubes, rods, or wire because of its tendency to seize in dies. Special lubricants have been developed to avoid the difficulty.

Many properties of zirconium are affected by the presence of contaminating elements. Hydrogen, oxygen, and nitrogen can all be absorbed by zirconium in huge amounts. At 1000 °C zirconium will absorb oxygen until its volume has increased visibly. 'Zircalloy' is an alloy of zirconium and lead, the latter added to reduce nitrogen absorption.

Zirconium has excellent resistance to nitric acid and withstands relatively high concentrations of other inorganic acids at room temperature. It is resistant to all but a few organic acids. The most important use of zirconium is in nuclear reactors for jackets for uranium fuel rods, for alloying with uranium, and for reactor core structures. A much smaller but important use is a powder form used as a 'getter' for cleaning up residual gases in vacuum tubes. Zirconium wire is often used as a flashbulb filler.

Because of its high resistance to corrosion, zirconium has found increasing use in the fabrication of pumps, valves, heat exchangers, filters, and other chemical handling equipment. As an alloying element, zirconium is important in the production of magnesium alloys and is used to a smaller degree in steels.

4
Joinability of Materials

Metal parts may be joined during manufacture by several processes:

1. adhesive bonding
2. soldering
3. welding
4. brazing
5. fastening

The first four methods are chemical processes wherein bonds are formed between clean surfaces by application of heat pressure, with or without fusion: the last-mentioned is a physical process.

The exposed surfaces of most metals are rough and may be contaminated with oxides, salts, and organic matter. The presence of these films and surface roughness inhibits bonding. In many joining processes, heat and pressure are applied to remove surface films and cause plastic deformation. However, if clean and highly smooth plane surfaces are brought together, cold bonding is possible.

Adhesives

Adhesives are natural and synthetic substances used for bonding or joining materials. Adhesives have been used for centuries for joining materials such as wood, paper, leather, rubber, and ceramics. The first types used were gums and waxes. These adhesives were derived from vegetable, animal, and mineral substances. Such adhesives provide low bond strength and are not suitable for the joining of metals. The so-called 'natural adhesives' are used principally in the wood and paper industries. Development of 'synthetic adhesives' which yield much stronger joints resulted in bonding being used in the aircraft industry during the Second World War.

Natural Adhesives

Natural adhesives are made from organic materials. There are four main groups:

1. Animal glues used in the paper and wood working industries and made from hides, hoofs, and bones.

2. Casein glues produced from milk have good adhesion properties to porous material such as wood, but lack moisture resistance.
3. Vegetable glues made from plant starches are cheap but lack strength; they are used on gummed stamps and envelopes.
4. Natural gums and resins, which the oldest type and have low melting points; some are used in the building industry, e.g., natural asphalt.

The above adhesives are inexpensive but lack strength and are susceptible to bacteria and fungal attack.

Synthetic Adhesives

Synthetic adhesives are plastic resins and may contain various other constituents. Curing agents cause the hardening of adhesives by chemical reaction and are available in liquid and powder form. Accelerators and extenders may be added to increase or decrease curing time. Chemically inert materials called 'fillers' may be added to modify viscosity, color, electrical properties, and strength. Synthetic adhesives may be divided into four groups:

1. thermosetting types
2. thermoplastics
3. elastomeric adhesives
4. anaerobics

Thermosetting Adhesives   Thermosetting types are usually more expensive but yield very high strength joints. They require heat and pressure during setting. The bonds produced are usually brittle and exhibit low impact strength. Main types of this group are epoxies, phenolics, ureas, and polyesters.

Epoxies have excellent adhesion to both porous and non-porous materials. They harden at room temperature. Typical bond shear strengths are 20 – 100 MPa. Good bond strengths up to 200 °C may be achieved.

Phenolics are used in the manufacture of aircraft and industrial honeycomb structures. Good adhesive strengths are achieved with porous materials. Phenol formaldehydes are used widely in plywood manufacture. Urea formaldehyde is a lower cost, commonly used adhesive; however, it is not waterproof and is prone to creep under load. Shear strengths on the order of 35 MPa may be obtained with ureas.

Polyester resins are used extensively as the binding matrix in glass fiber composites. Others are silicones, polyamides, polyurethanes, unsaturated polyesters, and melamines.

Thermoplastic Adhesives   Thermoplastics produce high strength bonds and are used for joining metals. They possess good impact strengths and good flexibility with joint shear strengths up to 20 MPa. Typically, thermoplastics are acrylics and cellulosics. Uses include transparent adhesives such as in automotive safety glass.

Elastomeric Adhesives   Elastomeric adhesives are based on synthetic rubbers such as nitriles and silicones. This group is the 'instant stick' type of adhesives which sets by solvent evaporation. Elastomerics generally do not produce very strong joints and are subject to creep.

Anaerobic Adhesives   In recent years the durability of acrylic polymers has been increasingly utilized in bonding engineering components. Adhesives described as anaerobic are compounds which polymerize spontaneously in the absence of oxygen, such as cyanoacrylates. Such materials must be packed in contact with atmospheric oxygen or they will cure during storage. The main uses of anaerobics are in securing non-sliding splines, locking threaded assemblies, and sealing gaskets.

A rapid growth area in adhesives has been that of pressure sensitive tapes. Some of the more specialized tapes are manufactured with lead or aluminum foil backings. Aluminum-backed adhesive tape is highly reflective to heat and light. A combination of aluminum foil laminated to glass cloth and a silicone adhesive forms a tape suitable for temperatures in excess of 316 °C. Uses of this specialized tape are in protecting electric cables, piping, and pump valves from heat.

Advantages of adhesive bonding are as follows:

> It is usually carried out at low temperatures, thus avoiding distortion.
> No holes are required, which permits the full strength of the member to be utilized.
> Joints yield a high fatigue strength due to uniform load distribution.
> Adhesive layers dampen vibrations.
> Galvanic corrosion is minimized since the adhesive serves as an insulator when joining dissimilar metals.
> Bonds can be made between different materials.
> Adhesives are often inexpensive and assembly costs can be reduced.
> Superior surface finish and closer tolerances on joints are often possible.

Joint quality and strength, however, are almost totally dependent on careful surface preparation and effective process control.

References

SEHGAL, S.P. and LINDBERG, R.A., *Materials: Their Properties and Fabrication*, 1973
Engineering Materials and Design, April 1972, IPC Industrial Press
Anaerobic adhesives and their applications, *Chartered Mechanical Engineer*, February 1976, Institution of Mechanical Engineers

Soldering

Soldering is defined as a joining process wherein coalescence between metal parts is produced by heating (usually below than 430 °C) and by using non-ferrous filler metals (solders) having melting temperatures below those of the base metals.

The union between solvent and base metal is a chemical bond produced by a metal solvent action. The solder dissolves a small amount of the base metal to form a solid solution. Upon solidification, the joint is held together by the same attraction between adjacent atoms that holds a single piece of metal together.

To achieve a sound soldered joint the following requirements should be considered: joint design, precleaning, fluxing, aligment, heat and solder, and flux residue treatment.

Joint Design

Solders as structural materials are weak in comparison to the metals which they are generally used to join; therefore, the soldered joint should be designed to avoid dependence on solder strength. The necessary strength can be provided by shaping the parts to be joined so that they engage or interlock, requiring the solder only to seal and stiffen the assembly. The two basic types of joints used in soldering are the lap joint and the butt joint.

Precleaning

Foreign material which is not wetted or alloyed interferes with the soldering process. An unclean surface will make soldering difficult if not impossible. All foreign materials such as oil, grease, paint, pencil markings, drawing or cutting lubricants, atmospheric dirt, oxide or rust films must be removed before soldering. The two general methods of cleaning are chemical and mechanical.

Chemical Degreasing, which involves the use of either solvents or alkalis to remove oil or grease from the surface, is often used. Acid cleaning or pickling is used to remove rust, scale, and oxides or sulfides from metal surfaces. Hydrochloric and sulfuric acid are used widely.

Mechanical Prepraration Several methods involving abrasion are most commonly used:

1. grit or shotblasting
2. mechanical sanding or grinding
3. filing or hand sanding
4. cleaning with steel wool
5. wire brushing or scraping with a knife or shaving tool

Flux

A soldering flux is a liquid, solid, or gaseous material which, when heated, is capable promoting or accelerating the wetting of metals with solder. An efficient flux removes tarnish films and oxides from the metal and solder and prevents reoxidation of the surfaces when heated. It is designed to lower the surface tension of the molten solder so that the solder will flow readily and adhere to the base metal.

Fluxes are classified into three main groups: highly corrosive, corrosive, and intermediate or noncorrosive fluxes. Good soldering practice requires selection of the mildest flux that will perform satisfactorily in a specific application. The highly corrosive fluxes consist of inorganic acids and salts. Corrosive fluxes are used to best advantage where conditions require a rapid and highly active fluxing action.

Intermediate fluxes are weaker than the inorganic salt types. They consist mainly of mild organic acids and bases and certain of their derivatives such as the hydrohalides. They are active at soldering temperatures, but the period of activity is short because of their susceptibility to thermal decomposition.

Water white resin dissolved in a suitable organic solvent is the closest approach to a noncorrosive flux. Its function as a flux is largely protective. Fluxes generally take the form of resins and pastes.

Joint Clearance

Clearance between the parts being joined should be such that the solder can be drawn into the space between them by capillary action. Usual joint clearances are about 0.08 – 0.12 mm.

Heating Methods

There are numerous methods for transmitting heat to the metal to be soldered: these include soldering irons, torches, dip soldering, induction and infrared heating, resistance heating, and oven heating.

The traditional soldering tool is the soldering iron with a copper bit which may be heated electrically or by oil, coke, or gas. Soldering irons are classified into four groups:

1. soldering irons for tradesmen
2. transformer-type low voltage pencil irons
3. special quick-heating type irons
4. heavy duty industrial irons

Regardless of heating method, the bit performs the following functions:

1. it stores and conducts heat from the source to the parts being soldered
2. it stores molten solder
3. it conveys molten solder
4. it withdraws surplus molten solder

The selection of a torch for soldering is controlled by the size, mass, and configuration of the assembly to be soldered. Where fast soldering is necessary,

54

a high flame temperature together with a large flame and large tip may be required. For slow heating the opposite is the case. Flame temperature is controlled by the nature of the gas or gases used. The highest flame temperatures are attainable with acetylene and lower temperatures with propane, butane, natural gas, and manufactured gas.

When conducted properly, dip soldering may be very economical inasmuch as an entire unit, comprising any number of joints, can be soldered merely by dipping the part in a heath of molten solder. It is necessary, however, to use jigs or fixtures to contain the unit and keep the proper joint clearance until solidification of the solder takes place. A recent application of dip soldering is the wave soldering technique which is used in making printed circuits.

The only requirement for a material that is to be induction heated is that it be an electrical conductor. Three types of equipment are available for induction heating: the vacuum tube oscillator, the resonant spark gap, and the motor-generator unit. Induction heating is generally applicable for soldering operations with the following requirements:

1. large-scale production;
2. application of heat to a localized area;
3. minimum oxidation of surface adjacent to the joint;
4. good appearance and consistently high joint quality;
5. simple joint design which lends itself to mechanization.

In resistance heating, the work to be soldered is placed either between a ground and a moveable electrode or between two moveable electrodes to complete an electrical circuit. The heat is applied to the joint both by the electrical resistance of the metal being soldered and by conduction from the electrode, which is usually carbon.

Oven heating is not widely used, but there are many applications where it has been found to produce consistent and satisfactory soldering. Oven heating should be considered under the following circumstances:

1. when entire assemblies can be brought to the soldering temperature without damage to any of the components;
2. when production is sufficiently great to allow expenditure for jigs and fixtures to hold the parts during the soldering;
3. when the assembly is complicated in nature and other heating methods are impractical.

Solders

Solders are usually alloys of lead and tin of varying composition. They form the actual bond between the metals being joined. Other solder compositions include alloys of:

1. tin, antimony, and lead
2. tin and antimony
3. tin and silver

4. tin and zinc
5. lead, silver, and tin

The solder composition used depends on the application.

Flux Residue Treatment

After the soldered joint is completed, flux residues that are liable to corrode the base metal or otherwise prove harmful to the joint effectiveness must be removed or made noncorrosive. The removal of flux residues is especially important where joints will be subjected to humid environments.

Oily or greasy flux paste residues are generally removed with an organic solvent. If resin residues must be removed, alcohol, petroleum spirits, turpentine, trichlorethylene, cyclohexanol, and most common organic solvents may be used.

Welding

As defined by The American Welding Society Handbook, a weld is a local coalescence of metal wherein coalescence is produced by heating to suitable temperatures, with or without the application of pressure, and with or without the use of filler metal. The filler metal either has a melting point about the same as the base metals or has a melting point below that of the base metals but above 425 °C.

From the wide range of welding processes available, only the more frequently used types will be outlined. These are electric arc, oxygen gas, electrical resistance, thermit, atomic hydrogen, inert gas metal arc, forge welding, and electron beam welding.

Electric Arc Welding

Metal arc welding consists essentially of a localized progressive melting and flowing together of adjacent edges of base metal parts by means of very high temperatures from a sustained electric arc between a metal electrode and the base metal. There are basically two types of electric arc welding:

1. Heat is supplied from the arc formed between a metal electrode (usually the same metal as the base) and the base metal. This electrode is consumed as the weld progresses.
2. Heat supplied between the arc from carbon electrodes or an electrode and the base metal. In this process the electrode is not consumed.

Voltages used in metal arc welding range from almost zero on short circuit to 40 V with a long arc. For carbon arc applications, somewhat higher voltages are used.

In the early days of welding uncoated electrodes were used, but coated

56

electrodes were introduced subsequently. The extruded coating on electrodes produces a gas shield which protects the molten pool of metal from detrimental effects of the atmosphere. They have the following advantages:

>increased welding speed
>more stable arc
>smoother weld beads
>improved mechanical properties

Principal advantages derived from arc welding are:

>high quality welds
>great flexibility
>high deposition rates
>low welding costs

Tanks, bridges, boilers, piping, structural machinery, furniture, ships and most commercial metals are joined by arc welding.

Oxygen Gas Welding

Oxygen gas welding employs a gas flame as a source of heat to raise the temperature of metal workpieces to their fusion points and thus allow the liquid bodies to flow together and solidify to form a bond.

Two basic gases are used in this process—oxygen and the fuel gas. Oxygen is required for combustion of the fuel. Hydrogen, acetylene, and various petroleum gases are among those suitable for welding, with acetylene being the most common. Atmospheric oxygen may give proper combustion in which case the flame obtained is referred to as the 'air-gas' type.

The gases are transported through hoses and a torch to the job. Gases are mixed in the torch and then passed through a tip to the flame. The size of the flame is determined by the size of the tip, and the larger the flame the more heat supplied. A welding rod of the same material as the parent metal is hand fed into the flame. The welding rod supplies filler material to form a continuous and homogeneous joint.

>Advantages: Easily controlled flame
>Very versatile and portable

>Disadvantages: Lengthy times may be required for pre-heating
>Possible harmful thermal effects to base metal
>Storage and cost

Oxy-gas methods can be applied to soldering, brazing, and to heating metal for working as well as welding. They are used extensively in sheet metal fabrication.

Electric Resistance Welding

If two metallic materials to be welded are placed between two low resistance conductors (electrodes) and a low voltage, high current source of electricity is applied, the materials will be heated because of their electrical resistance. If some means is provided for forcing the pieces together after being thus heated to a plastic temperature, a bond is created at the contacting surfaces between the electrodes. To complete a weld satisfactorily, the current must be terminated and pressure retained until the weld metal strengthens after cooling. The metallurgical advantage of this process is that the metal is held at a temperature that is within the grain-growth range for only a short time. Electric resistance welding may exist in several forms such as spot welding, seam welding, projection welding, flash welding, and upset welding.

Spot Welding Most spot welders are of the stationary type, but a high degree of maneuverability is obtained with spot welding guns which are powered and actuated through flexible leads and hoses.

The electrodes which conduct current and which subsequently apply pressure in spot welding are usually low resistance, hard copper alloys. Electrode materials should be of high thermal conductivity to remove heat rapidly from areas of contact with the workpiece.

Resistance between the electrodes and the workpiece should be kept low for satisfactory spot welding. Metal surface cleaning may be required to remove dirt and greases. In order that the weld section be symmetrical about the weld interface, both workpieces must be in contact, be of equal thickness, and be of the same material. Spot welding is used widely in the automotive industry and in light steel construction.

Seam Welding The principles of seam welding are the same as for spot welding, the chief difference being that wheel-type electrodes are used instead of cylindrical electrodes. In the seam welder, a series of spot welds are made by passing the workpiece between the electrodes, giving rise to a continuous 'seam' of weld, rather than a series of discrete 'spots'. Electrode pressure is fairly constant throughout the seam length and the welding current may be regularly interrupted to produce the 'chain spot' effect. Spots are so placed that they overlap and thus form a continuous, fluid-tight weld, as can be seen in fuel tanks, barrels, and mufflers. Instead of spots being round, as in spot welding, they are usually oblong or rectangular. The process is sometimes called 'stitch welding'.

Use of seam welders is restricted generally to sheet metal fabrication because of equipment cost and electric supply difficulty with large units.

Projection Welding In projection welding, current and heat during welding are localized at predetermined points by means of special shapes of the parts to be joined, e.g., projections or knobs are provided on the welding surfaces. Spot welds result; however, electrode rods are not required since the projections serve the same purpose. Thus machine capacity is increased appreciably by welding several spots or projections simultaneously. Projection welding is most advantageous in assembling parts made by punching or stamping.

Materials which may be projection welded are somewhat limited since certain metals do not have sufficient strength. Copper, some brasses, and some aluminum alloys are considered unweldable by the process. Most wrought ferrous materials lend themselves readily to projection welding.

58

Flash Welding   Flash welding is achieved by placing in light contact two pieces of metal which are to be welded and passing sufficiently high current through them to cause arcing. Once fusion temperature is reached, pressure is applied to force the fused areas together. Flash welding is usually restricted to welding the end of one piece to that of another, both having the same cross-sectional area. Applications include end-welding strips, bars, and rims.

Most commercial metals may be flash welded. Some exceptions are antimony, bismuth, lead, zinc, and many alloys in which those elements are present. Cast iron may be flash welded but it tends to form brittle martensite. All types of steels have been satisfactorily welded to their own type, to other types, and to most other metals except those noted above.

Upset Welding   Upset welding—similarly to flash welding—consists of placing two metal pieces in contact, passing a current through them, and applying pressure to force the surfaces together. Upset welding is unlike flash welding in that heat is generated due to the inherent resistance of the materials under greater contact loads. Flash welding is usually preferred over upset welding because:

> greater production speed results
> power demand is more favorable
> no special preparation of weld surfaces is required
> better mechanical properties are obtained

Thermit Welding

The thermit welding process is classified as nonpressure welding. High temperatures obtained from the reaction of finely divided iron oxide and aluminum are employed to raise the temperature of parts to be welded above their fusion point according to the reaction

$$8Al + 3Fe_3O_4 \rightarrow 9Fe + 4Al_2O_3 + Heat$$

Thus when finely divided aluminum and iron oxide react, the products are pure iron, aluminum oxide, and great amounts of heat. Temperatures obtained from the reaction are about $2500 - 2800$ °C.

Thermit mixtures are ignited by special ignition powder consisting largely of barium peroxide. Thermit welding is used mainly for the repair of shafts, machinery frames, gears, and railroad rails. Success of joints in such applications depends on alignment of the workpieces.

Atomic Hydrogen Welding

Atomic hydrogen welding is also a non-pressure fusion process, however, the heat source is obtained by recombination of dissociated hydrogen. The molecular form of hydrogen is broken down into the atomic form by passing the gas through an electric arc and, upon its emerging from the arc, recombining to the molecular form and giving off heat.

The electric arc in atomic hydrogen welding differs from ordinary welding arcs because it is sustained between two permanent tungsten electrodes rather than between an electrode and the parent metal. Thus, there is somewhat better heat control than in the metallic arc-welding processes. The arc and weld pool are completely surrounded at all times by an envelope of hydrogen gas and flame, so that it not only heats the work but also provides shielding from impurities.

Filler rods used in atomic hydrogen welding are generally the same as the parent metal composition and are usually added by melting off in the flame tip. The atomic hydrogen process is used on practically all metals and alloys as are welded by the metallic arc and the oxyacetylene processes. Flux is not usually required except for aluminum and copper and their alloys.

Metal Inert Gas (MIG) and Tungsten Inert Gas (TIG) Welding

MIG welding makes use of a consumable metal electrode. The arc stream and molten metal are protected by an envelope of inert gas such as helium or argon. Some metals such as zinc alloys and rimmed steel give off fumes and gases, contaminate the protective envelope, and produce high porosity in the weld deposit.

In usual practice, argon gas is used with a.c. welding whereas helium gas is used with d.c. welding. The process is applied to most metals, although generally it is used on only the hard-to-weld metals such as aluminum, stainless steel, and others where its cost is justified. Carbon dioxide is another gas frequently used in shielded welding.

A somewhat similar process to MIG welding is TIG—tungsten inert gas—welding. The later also uses argon or helium gas; however, a non-consumable tungsten electrode is used. TIG welding may be used to join aluminum, columbium, magnesium, and molybdenum.

Forge Welding

Classified as a pressure weld, forge welding employs heating of the pieces to be joined in a forge fire to a highly plastic state and completing the weld by applying pressure. Pressure application is usually in the form of hammer impact over an anvil.

The forge fire is generally oxidizing and workpieces heated therein would have oxide films at welding temperatures unless a flux such as borax is sprinkled on the welding surface. Temperatures required for forge welding vary with the materials being welded. Joints suitable for forge welding are the lap, scarf, butt, and Vee. Forge welding does not have widespread use except in agricultural communities and in some heavy equipment shops. Other processes are faster, less expensive, and more adaptable for most work.

60

Electron Beam Welding

Electron beam welding employs a heated filament which produces a supply of free electrons. A high voltage source is connected between a gun and workpiece which accelerates electrons to a velocity half that of sound. These high-impact velocities produce the heat for welding. No filler rod is used. The purpose of the focusing magnetic coil employed between gun and workpiece is to concentrate the stream of electrons into a narrow beam.

In electron beam welding only a very narrow fusion zone is produced and only the welded joint is heated. Thus, there are no thermal effects on the parent metal. However, this type of welding must be performed in a vacuum and X-rays are also produced. 'EB' welding is used to join and cut refractory metals.

Recent Advances

Recent developments in welding have extended into the use of niobium strengthened compositions, use of greater weld thicknesses, and to more severe climatic regions—the Arctic for example. In addition, steel rolling and heat treating practices have become more sophisticated and many evaluations are now concerned with all regions of the weld and adjacent areas.

Recent technology has also examined welding of HSLA steels, concentrating mostly on microstructural alterations of the ferrite – cementite phase with significant changes in toughness in weld properties.

Improved discontinuity detection using computer-aided ultrasonic pulse – echo techniques is being used. The approach offers a means of obtaining improved characterization of the size, shape, and location of subsurface discontinuities. Computerized data processing techniques are used on the signal obtained in conventional ultrasonic systems; principal benefits are improved signal-to-noise ratio and resolution.

Criticism has been made of fume generation in arc welding processes, which must lead to methods for reducing it. Composition analysis of the fumes generated by such welding processes indicates that fumes are produced by two mechanisms. The first is simply vaporization of the metal or compound from the vicinity of the arc and subsequent condensation and oxidation. The second is greater vaporization due to the formation of a more volatile oxide on the surface of the molten electrode, followed by condensation and perhaps further oxidation. The degree to which each of these mechanisms participates in fume production, composition, and rate of fume evolution depends on the process used.

Classification of Weld Defects

Weld defects may be broadly classified as in the following outline:

1. Defects involving inadequate bonding
   (a) Interference weakening in pressure welds
   (b) Incomplete penetration in fusion welds

2. Foreign inclusions
   (a) Oxide films in fusion welds
   (b) Slag inclusions
   (c) Delaminations
   (d) Tungsten inclusions
3. Geometric defects
   (a) Undercutting
   (b) Excessive reinforcement
4. Metallurgical defects
   (a) Defects related to microsegregation
       (i) Hot cracking and microfissures
       (ii) Cold cracking and delayed cracking
       (iii) Stress-relief cracking
       (iv) Strain-age cracking
       (v) Gas porosity
   (b) Problems arising from metallurgical reactions
       (i) Embrittlement
       (ii) Structural notches

References

Welding Journal, January 1975, July 1975, July 1976, American Welding Society

Welding Production, Volume 19, Number 7, July 1972, The Welding Institute, Miami

Welding Handbook, American Welding Society

PLUGER, A.R. and LEWIS, R.E. *Weld Imperfections*, Addison-Wesley, Reading, Mass., 1968

MORRIS, J.L. *Welding Processes and Procedures*, Prentice-Hall, Englewood Cliffs, N.J., 1965

Brazing

Brazing is defined as a joining process wherein coalescence is produced between metals by heating them to a temperature above 425 °C and by using a nonferrous alloy which has a melting point below that of the base metal. In brazing, the parent metals are not fused but, since the temperature is high, appreciable diffusion and alloying is possible between the brazing alloy and parent metals. The main feature of brazing is that many dissimilar metals can be readily joined, as well as cast and wrought types. The filler strength in brazing is usually less than the base metals.

Brazing is used on aluminum and its alloys, alloys of nickel and copper, and ferrous metals including carbon and low alloy steels, stainless steel, high speed steel, and cast iron. It also finds use among the more rare metals such as molybdenum, titanium, tungsten, and zirconium.

The brazing alloys are usually copper, brazing brass, phosphor-copper,

copper – silver alloy, nickel – silver, silver alloy, copper – silicon alloy and aluminum – silicon alloy. The usual flux for brazing of steel is moisture free borax made into a paste in alcohol while other compositions are made up of chlorides and fluorides of the alkaline earth metals.

The process of brazing is achieved by heating the metals in the joint area by a torch, a furnace or an induction coil, to a temperature above the melting point of the filler material. Brazing results in a moderate strength bond, greater than that obtainable with soldering and usually less than that available with welding.

Mechanical Fasteners

The basic function of mechanical fasteners is to provide an adequate economic clamping force between two pieces of material under all foreseeable conditions of loading. With the advent of thousands of different types of fasteners such as screws, rivets, nuts and bolts, nails, and modern quick release fasteners, the selection of proper fastenings is difficult.

Fasteners are available in a wide range of engineering materials. Selection is normally based on such considerations as environment (i.e., corrosive or temperature extremes), weight, magnetic properties, stresses, reusability, and life expectancy, as well as cost.

For the greatest economy, standard materials should be used. Specifying a fastener material with a given chemical analysis adds time and cost. Often a standard fastener can be altered by heat treatment, cold working, or coating to meet special needs. Most fasteners are made from steel. Specifications cover a broad range of mechanical properties that are indicated by a bolt-head marking system that identifies the fastener by grade.

The family of aluminum alloys is the least costly by volume of all fastener metals. Aluminum fasteners are classified as hardenable and nonhardenable and weigh about one-third as much as steel. Some grades equal or exceed the strength of mild steel. The metal polishes to a high luster, has high thermal and electrical conductivity, is non-magnetic, can be hardened by alloying, and has high corrosion resistance.

Brass is easily worked and has moderate strength. One of the most malleable of all metals, copper, can be worked into a wide variety of shapes. It has good corrosion resistance and the highest conductivity of all nonprecious metals.

Fasteners can be made from commercially pure nickel, Monel or Inconel. They are typically used where toughness, immunity to discoloration and corrosion, and strength at high temperature are desired.

Stainless steel fasteners are used where corrosion, temperature, and strength are problems. They also have a mirror-like finish. Lighweight, high-strength fasteners made from titanium are used chiefly in aircraft as a weight saving measure. Titanium fasteners are most commonly used in joints loaded in shear but are also used in tension-loaded joints.

Exceptionally lightweight fasteners of beryllium are about 40% as heavy as the equivalent titanium fasteners. Brittleness is still a major limitation to widespread use. Beryllium bolts are used primarily for applications with high shear loads.

Plastic materials can meet a broad spectrum of design requirements—optical, strength, rigidity or flexibility, heat, low temperature, chemical, and corrosion resistance, sealing, toughness, good electrical properties, and light weight. Common plastic fastener materials are often less costly than their metal counterparts. In addition to being available in standard fastener shapes, plastic fasteners can be designed to have special functions. Integral color, special undercuts, and molded-in metal inserts, are a few of the possibilities. Assembly can often be simplified when a single plastic fastener replaces several metal components.

The types of fasteners may be considered under the following headings:

1. Threaded fasteners
2. Non-threaded fasteners
3. Special-purpose fasteners

Threaded Fasteners

Threaded fasteners remain the basic industrial assembly method despite advances in welding, adhesives, and other joining techniques. Over 500 000 standard fasteners can be classified, i.e., their physical characteristics are covered by published standards. At least as many 'specials' are also made. These fasteners are engineered for specific product needs.

A bolt is defined as a threaded fastener intended to be mated with a nut, while a screw engages either preformed or self-made internal threads. Bolts and screws share a number of head shapes and drive configurations. Only those bolts which have a shank configuration preventing rotation cannot be considered—or used—as a screw. Conversely, only those fasteners with cutting or tapered threads not intended to mate with a nut, such as self-tapping screws, cannot be used as bolts.

A stud is an externally threaded headless fastener. One end usually mates with a tapped component and the other with a standard nut: Tapping screws cut or form a mating thread when driven into preformed holes. These one-piece fasteners permit rapid installation since nuts are not used and access is required from only one side of the joint—an important consideration in design. The mating thread produced by a tapping screw fits the screw threads closely and no clearance is necessary. This close fit usually keeps the screw tight, even when subjected to vibration. Tapping screws are usually case-hardened and have tensile strengths of 700 MPa or greater with relatively high ultimate torsional strengths. Tapping screws are used in steel, aluminum die-castings, cast iron, forgings, plastics, reinforced plastics, asbestos, and resin-impregnated plywood.

Externally threaded fasteners with locking devices use the same techniques to achieve locking as locknuts. Prevailing-torque locking screws employ distorted threads, interference fits, inserts, or a chemical adhesive coating. Free-spinning locking screws and bolts have a locking feature incorporated into the head or an additive on the threads.

Studs require a two-part assembly operation but offer several distinct advantages. Studs eliminate the problem of deviations from perfect squareness

64

in an assembly. The ability of a nut to 'float' and adjust on the nut end threads is one main advantage of using a stud over a bolt or screw. In assembling and reassembling heavy parts such as turbine casings and cylinder heads, studs can act as pilots. In the automatic assembly of small lightweight units, studs reduce assembly cost since they permit quick and easy 'stack-up' of gaskets or other different parts of a joint. Studs reduce the need for the large hole clearance and close hole alignments required by a cap screw or bolt. Studs with an interference-fit thread or lock-thread on the tap-end provide a positive lock against turning and loosening. The lock facilitates assembly and disassembly of lock-nuts on studs and is particularly important where maintenance of prestress is required to combat fatigue failure. Studs supplied with sealant prevent leakage of fluids through holes tapped in porous materials.

Nuts are internally threaded fastener elements designed to mate with a bolt. Hexagonal and square nuts, also called full nuts, are the most commonly used. Square nuts are normally used for lighter duty than hexagonal nuts. Surfaces may be flat, chamfered, or washer-faced. Flanged nuts are available with integral washers to simplify handling and may be used on oversize holes. Under many assembly or load conditions, standard hexagonal or square nuts cannot maintain a sufficiently tight joint.

Non-threaded Fasteners

Non-threaded fasteners include rivets, retaining rings, nails, stitching and stapling, pins, and washers.

Rivets are low-cost, permanent fasteners. They are well suited to automatic assembly operations. The primary reason for riveting is low in-place cost. Initial cost of rivets is substantially lower than that of threaded fasteners because rivets are made in large volumes in high-speed machines with little scrap loss. Assembly costs are low since rivets are usually clinched in place by high speed machines. Another advantage is that dissimilar materials, metallic or non-metallic, in various thickness, can be joined providing problems of electro-chemical corrosion are avoided. Almost any part can be fastened by a rivet provided there are flat parallel surfaces for the rivet clinch and head and adequate space for the rivet driver during clinching.

Tensile and fatigue strengths of rivets are usually lower than for comparable bolts or screws. High tensile loads may pull out the clinch and severe vibrations may loosen the fastening. Riveted joints are neither watertight nor airtight, although such a joint may be attained by using a sealing compound. Riveted parts cannot be disassembled for maintenance or replacement without destroying the rivet.

Retaining rings, also called snap rings, provide a removable shoulder to accurately locate, retain, or lock components on shafts or in bores and housings. They are usually made of spring steel and have high shear strength and impact capacity. Some rings are designed for absorbing end-play caused by accumulated tolerances or wear in the parts being retained.

Nails are the simplest of fasteners and are used extensively in woodworking, building construction, and the packaging fields.

Stitching is a fastening method in which U-shaped stitches are formed from a continuous length coil of steel wire by a machine which also applies the stitch to the materials being joined. The advantages of stitching are:

dissimilar materials can be joined
fastening takes place at high speed
hole prepunching is eliminated
minimization of disturbances to coated materials

Stapling differs from stitching in that individual staples are preformed and cohered into strips.

Pins are inexpensive and effective fasteners for use where loading is primarily in shear. They are divided into two groups: (a) semi-permanent pins which require the application of pressure or the aid of tools for installation or removal; (b) quick-release pins which are used where parts are frequently dismantled and assembled.

Washers are added to bolts and screws to serve one or more of the following functions:

distributing the load
locking the fastener
protecting the surface
insulating
spanning oversize holes
sealing
providing electrical connections
providing spring tension

Special-purpose Fasteners

Special fasteners include threaded and non-threaded types and fastening elements such as sleeves and latches designed for specific needs. They frequently incorporate the functions of several pieces, thereby simplifying assembly. Cost savings attainable with special purpose fasteners can be substantial, particularly in labor and materials.

References

ASTM, *Plastics Tooling and Manufacturing Handbook*, Prentice-Hall, Englewood Cliffs, N.J., 1965

CLEMENTS, R., *Manual of Light Production Engineering*, Business Books, London, 1968

Machine Design reference issue 'Fastening and Joining', 20 November 1975, Penton Publishing Co.

SOLED, J., *Fasteners Handbook*, Reinhold, London, 1957

5
Electrical and High Temperature Materials

Carbon and Graphite

Carbon exists in two crystalline forms, diamond and graphite, both of which are found naturally and can be produced industrially. Diamond is the hardest substance found in nature, with a value of 10 on the Moh scale. Pure diamonds are colorless and transparent but, due to impurities, naturally occurring diamonds are usually colored red, blue, green, or yellow. Graphite is widely distributed in nature and is found as a soft, gray-black, shiny material. It is also manufactured.

Besides crystalline forms of carbon, there are noncrystalline or amorphous carbons such as carbon black, lampblack, coal, coke, and charcoal. These different forms of carbon, although chemically identical, vary in hardness, specific gravity, and other physical properties.

General Properties

At ordinary temperatures and pressures all forms of carbon are stable, but they normally revert to graphite at high temperature. Diamond can be transformed to graphite at temperatures of 1700 – 1900 °C. Artificial graphite can be produced from amorphous carbon, for instance from petroleum coke, by heating in an electric furnace at about 3000 °C. The transformation of graphite to high quality diamond at high pressures appears theoretically feasible, but so far has not been achieved. Industrial diamonds of lower quality are, however, made in this manner.

Graphite and carbon are excellent refractory materials, but they can only be used under neutral or reducing conditions because they oxidize readily in air at elevated temperatures. At very high temperatures, graphite is a more stable form than amorphous carbon, which is converted to crystalline graphite on prolonged heating at about 2500 °C. Graphite has no melting point but sublimes at a temperature of 4200 °C. Graphite is a valuable refractory in high temperature applications such as rocket nozzles and nozzle throat inserts. Many attempts have been made to improve the oxidation resistance of graphite by applying coatings of silicon carbide or molybdenum silicide or by impregnating its surface with molten metals such as zirconium. Zirconium carbide thus formed provides a good chemical bond with the graphite surface, whereas the outer layer of zirconium is oxidized to zirconia, thereby protecting the graphite base from oxidation.

Carbon and graphite are highly resistant to acids, alkalis, and all solvents except oxidizing ones. In metal-cleaning operations, carbon has the dubious honor of being the most difficult type of scale to remove from metal surfaces since standard chemical metal cleaners will not remove it.

The thermal conductivity of carbon, particularly of graphite, is much higher than that of other nonmetallic materials and approaches the conductivity of metals. The tensile strength of carbon materials varies with the type, grade, and method of manufacture and ranges from 8 to 25 MPa at room temperature. The tensile strength approximately doubles at temperatures above 2460 °C. The spall resistance of carbon materials is so outstanding that it is difficult to devise an experimental method of cracking them by thermal stresses alone. Carbon and graphite also possess low thermal expansion and low neutron absorption cross-section.

Graphite is more easily machined than other forms of carbon. Graphite is produced by heating carbon black to 2760 °C or, if heated to 1510 °C, baked carbon is produced. Baked carbon is amorphous and extremely hard and has lower conductivity than graphite. Carbon and graphite shapes manufactured by usual techniques are porous. To make them impervious to fluids, they are impregnated with a synthetic resin, usually a phenolic type. The impregnated material is slightly permeable, but when the resin is polymerized the product becomes impervious to fluids under pressure and its strength increases. Both industrial carbon and graphite have low densities.

Graphite has a lower specific resistance than other phases of carbon. It is therefore preferred as an electrical conductor despite its higher cost. By far the greatest tonnage of carbon and graphite is used in the metallurgical and electro-chemical industries as anodes in arc-melting furnaces.

Uses

Carbon in Steel   Carbon is the most important alloying element in steel. It is very potent in its effect on steel characteristics and few steels contain more than 1% carbon. Those steels in which carbon is the only significant alloying element are termed carbon steels. Some effects of increasing carbon content in steels are:

    increased hardness
    increased wear resistance
    increased yield and ultimate tensile strength
    more difficult weldability
    more expensive
    reduced ductility and impact strength
    reduced machinability
    reduced melting point
    steel becomes heat treatable

There is virtually no change in the modulus of elasticity of most steels as carbon content is increased.

68

Baked and Graphitized Products   The term 'baked carbon' refers to products made by baking mixtures of carbonaceous raw materials (i.e., petroleum, coke, and coal) with a binder such as tar or pitch at about 1000 – 1800 °C. Many baked products consisting of amorphous carbon are widely used, but they may be submitted to further heat treatments at above 2000 °C to form graphitized products or electrographite. The principal uses of baked and graphitized carbon products are as:

> anodes used in electrolytic industries
> brushes for electric generators
> electrodes for the aluminum industry
> electrodes used in the production of calcium carbide, phosphorus, and many ferro-alloys
> electrodes used for stainless steel-producing furnaces

Carbon Isotopes   Carbon 13 is one of the isotopes of carbon and is used as a tracer in biological research because its heavy density is easily distinguished. The isotope carbon 14 is widely used in archeological dating. Carbon 14 or radioactive carbon exists in air and is formed by the bombardment of nitrogen by cosmic rays at high altitudes.

Carbon Fibers   Carbon fibers are made by the pyrolysis of precursor fibers such as polyacrylonitrile or other long chain molecule materials. The carbon has a crystalline form and the fibers are strong, have high modulus, and are flexible. Carbon fibers are produced as continuous filaments with strengths comparable to high strength steels. Conductive fabrics are also made from carbon yarns.

Carbon Brick   Carbon brick, used as a lining in the chemical processing industries, is carbon compressed with a bituminous binder and then carbonized by sintering. If the binder is capable of being completely carbonized, the bricks are impervious and dense. Graphite brick, made similarly from graphite, is more resistant to oxidation than carbon brick, has a higher conductivity, and is softer. Impervious carbon in used for lining pumps, for valves, and for acid resistant parts.

Pyrolytic Graphite   Pyrolytic graphite is made by a gas or vapor plating process. Microscopic flaws are thereby eliminated. A special feature of pyrolytic graphite is its strong orientation of crystal axes and hence high degree of asymmetry of physical and mechanical properties. Uses of the material include rocket nozzles, nose cones, and various other nuclear, chemical, and metallurgical applications.

Recrystallized Graphite   By a hot working process, recrystallized or 'densified' graphite can be produced with bulk densities in the range $1.85 - 2.15 \times 10^6$ g/m^3 as compared with the range $1.5 - 1.7 \times 10^6$ g/m^3 for conventional graphites. Advantages are:

> ability to take surface finish
> absence of structural micro-flaws

69

high quality reproductivity
improved creep resistance
lower permeability

References

PATTON, W.J. *Materials in Industry*, Prentice-Hall, Englewood Cliffs, N.J.,' 1968

ABBOT, H.W. Carbon—baked and graphitized, reprinted in *Encyclopedia of Chemical Technology*, Vol. III (R.E. Kirk and D.F. Othmer, eds), Interscience Publishers, New York, 1949

Engineering Materials and Design, Volumes 18, 10, 20, 16, 1974–75, IPC Industrial Press, London

Electrical World, 10 March 1975, Strand Publishing Pty Ltd,

Laminated Thermosets and Glass Fiber

The development in recent years of high modulus and high strength boron and graphite fibers bonded together in plastic or metal binders has brough a renewed interest in layered structural materials. Typically these materials involve multiple layers of fibers with each layer treated as a homogeneous anisotropic layer.

Laminated Thermosets

A laminated material is composed of a stack of fibrous or porous sheets impregnated or coated with a resin or binder bonded together by heat and pressure to form an integral body. The plastics industry manufactures large quantities of laminated sheets, tubes, rod, and products using various materials such as asbestos, cloth, glass fabric, paper, wood cellulose, etc., bonded by synthetic resins. Laminates are divided into groups according to the bonding pressure required. Those laminates cured above 7 MPa are referred to as high pressure; those below 3 MPa as low pressure; and those below 0.2 MPa as contact pressure.

Thermoset laminates are of the high pressure group. They are the hardest and stiffest of all plastics and are chemically insoluble. They are cured by chemical reaction causing crosslinking of the polymer chains. Some resins most commonly used are:

epoxies
melamines
phenolics
polyimides
silicones

The resins are discussed in some detail as they give the lamina its particular characteristics.

70

Epoxy resins are expensive, but provide a high degree of resistance to acids, alkalis, and solvents, along with low moisture absorption and high structural properties.

Melamine resins are particularly ideal for domestic application. They exist in many colors and can carry designs such as wood grains or geometric patterns, and their surfaces retain initial polish and are easily cleaned. They are unharmed by alcohol, grease, oils, or soap and are only slightly harmed by cigarette burning. They have good mechanical strength and shock resistance.

Glass-reinforced phenolic laminates are well suited for structural use. The material has better high temperature strength than aluminum. In general phenolic laminates show good mechanical and electrical properties, flame resistance, and low cost. They are used widely in electrical applications and in aircraft and missiles.

Polyimides are usually applied to glass cloth, boron, and graphite fabric laminates. They retain almost full strength at temperatures to 375 °C which makes them well suited for aerospace applicaitons. Very high pressures are required in their manufacture.

Cured at high temperatures, silicone resin laminates are high in cost and provide outstanding heat resistance.

Uses Use of laminated thermosets is extensive because of the great number of laminate – resin combinations. They can, however, be classified according to decorative, mechanical, and electrical types. Decorative laminates are used as table tops, shelves, counters, name plates, etc., because of their surface finish and their wear and stain resistance. Mechanical laminates are used in quite, heavy duty gears because of their resilience. Cam followers, bearings, and clutches often employ laminates. Most laminates are mechanically nonresonant and tend to dampen and reduce sound vibration. Copper clad laminates are used for printed circuitry and in high voltage applications.

Glass Fiber

Glass fibres were first made about 1500 BC and were used for decorative purposes. Much development since 1942 has led to glass fiber reinforcement becoming a significant engineering material.

The basic continuous glass filament is produced by heating glass in a platinum crucible until it is molten. It is then swiftly drawn through orifices and collected. Staple length fibers can also be formed by blowing compressed air across a molten glass filament stream.

A starch – oil binder (called a 'size') is applied to minimize loss of strength in handling the fibers. A 'finish' may also be applied to promote subsequent bonding of the fiber to a resin. Continuous fibers are wound on high speed collector reels.

Types of Glass Fiber There are six types of glass fiber. 'E' glass is the major proportion of all glass produced. It is a lime – alumina – borosilicate composition having high bulk electrical resistance, high surface resistivity, and good fiber forming characteristics. 'A' glass is a soda-lime glass. Its properties are inferior to E glass but it is cheaper to produce. 'C' glass has very good acid resistance. 'D' glass has a low dielectric constant. 'M' glass is noted for its hardness and modulus. 'S' glass has high strength, modulus and high strength retention at elevated temperature.

A recent product marketed by Pilkington in the UK is an alkali resistant glass fiber known as Cem-fil which is used as a cement reinforcement.

Product Types   Glass fiber is available in a variety of forms.

1. *Yarns* are produced for use in woven fabric manufacture. Several strands of fiber are twisted together and then doubled over.
2. *Woven fabrics and tapes* are widely used in electrical applications and as high temperature protective insulation on exposed engine parts and exhausts.
3. *Rovings* are several continuous strands bundled or collimated which result in high tensile strength. Rovings are ideally suited for producing fishing rods, racquet shafts, and are used in filament winding.
4. *Chopped strands and strand mat* are made from continuous strands cut into short lengths and bound together randomly. Mats of continuous fibers are also produced. Glass fibers are manufactured into cloths for reinforcing and for nonstructural applications. Cloths are also used in other forms, e.g., wood for insulation.
5. Tissue and small particles, almost powders, are made for particulate composite material fillers.

Properties and Applications   As a reinforcement material, glass fibers have marked advantages over other reinforcements such as cotton or asbestos. Some advantages are:

    complete fireproofness
    general chemical inertness
    good weathering properties
    high modulus-to-weight ratio
    high strength-to-weight ratio
    low extensibility

Manufacturers of electrical equipment and wire and cable are major users of glass fiber textile products. Resistance to chemicals, moisture, rot, and high temperatures combined with high strength/low bulk ratio make these products an excellent structural medium and reinforcement for impregnants used for insulation in electrical equipment.

The chief use of glass fiber is for reinforcements. Of these, woven cloths are the most expensive (on a weight basis) but they have the highest values of physical properties. Cloth is used where maximum properties are required and cost is of secondary importance (i.e., aircraft). Mats and rovings are generally employed where cost is the prime criterion (i.e., boats, caravans, translucent sheeting).

Glass fiber reinforcement results in moldings with physical properties superior to those associated normally with plastics. Some benefits of glass fibers reinforced plastics are that they

- can be formed into almost any shape, simple or complex, large or small
- can deliver more strength per unit weight than any unreinforced plastic and most metals
- can be designed to provide an wide range of tensile, flexural, and impact properties
- have outstanding electrical properties
- hold their shapes under severe mechanical and enviromental stresses
- eliminate the necessity for painting because color can be molded into the laminate, for long lasting good appearance
- will not rust or corrode and will provide resistance to almost every chemical environment

Water absorption, resin crazing, and smoke/toxic fume emission in fire conditions can be problems not encountered with most metals.

Uses for glass fiber reinforced plastics (GRP) include agricultural equipment, appliances, automobile body parts, aviation and aerospace. Other fields where GRP is used extensively are business machines, chemical processing, construction, electrical/electronics, house and home, marine, materials handling, recreational, and transportation. GRP can be manufactured by a variety of methods:

centrifugal casting
contact molding
cold molding
continuous laminating
hand or machine lay up
filament winding
matched molding

Mica

Natural mica comprises a series of minerals of which two, muscovite and phlogopite, have important commercial value. Mica crystal can be split into thin lamina which possess insulating and heat resisting properties and are highly elastic perpendicular to their surfaces. Mica is found in all parts of the world but commercially important deposits occur in India, Madagascar, the USA, Canada, South America, South Africa, and the USSR.

Mica block and film are graded into standard sizes, each of which is designated by a number based on the largest rectangular area the block will produce. Mica can be further classified by visual inspection on the basis of type and degree of staining, air inclusions, waviness, cracks, and color. Splittings are size-graded by a similar number system but their quality depends on uniformity of thickness, regularity and soundness of edges, and type and extent of staining.

Ground Mica

Scrap mica from mining and factory waste and schist mica are ground to a fine powder by both wet and dry processes. The major uses are in roofing compositions, paints, and rubber compounding. Miscellaneous uses include fillers in molded-plastic and glass compositions, christmas-tree 'snow', oil-well drilling fluids, axle grease, and oils. In these applications various properties of small mica flakes such as electrical and heat resistance, non-flammability, lubricity, and tendency to leaf are advantageous.

Mica Splittings

These are manufactured into 'pasted mica' for electrical machines. Pasted mica is made by bonding individual splittings with natural or synthetic resins to form hard plates or flexible sheets. In many cases reinforcing materials such as papers, cotton or glass cloth, plastic film, or asbestos are combined with mica to add tensile strength, tear resistance, or other desired properties. These materials fall into several types.

Segment Plate A hard mica used for separation of the segments or bars of commutators. Both muscovite and phlogopite splittings are used. The thermosetting bonding resins used are fully cured in making the plate to prevent slippage or shear between splittings during commutator operation under load. Elasticity is needed to absorb the expansion and contraction of the copper bars under temperature and load changes.

Molding Plate Similar to segment mica but is made only from muscovite and usually has higher binder content. The bonding resin is only partly cured in pressing so that under subsequent heat and pressure it can be formed into various shapes such as rings and tubes.

Cold Forming Flexible Plate Uses a permanent thermoplastic bonding resin and may be made on a continuous machine, from loose splittings, or by hand. Individual book form splittings are laid by hand on screens with the bonding resin brushed between layers. To impart high tear resistance, mica is laid on glass cloth or on paper. The material is sheared to size for insulation in small motors and generators.

Tapes and Wrappers Insulate armature coils in rotating machines. Wrappers are made generally by hand laying splittings on a continuously moving sheet of reinforcing material and dropping or brushing a solution of thermoplastic resin between layers. High quality wrappers, about 0.25 mm thick, are usually proof tested at 3000 – 5000 V during manufacture to detect weak spots caused by poor splittings or improper laying. Most tapes are made by slitting a wide sheet, hand built in the same way as a wrapper, to the desired width.

Heater Mica Made by hand-scattering small, loose, amber or white splittings on a wire screen using a very low binder content. It must always be supported since the binder burns away in a short time. Applications include percolators and similar appliances such as toasters which operate below 800 °C.

Mica – Asbestos Combinations Made by sandwiching flexible or molding

74

mica between sheets of high-grade asbestos paper. For continuous service over many years, pasted mica insulation in machines is usually not stressed at above 2400 – 3200 V/mm.

Block and Film Mica

The highest quality muscovite block mica is used in radio condensers operating at many frequencies and voltages. The most important property in this application is mica's low power factor. Also of importance are high dielectric strength and insulation reistance, stability under operating conditions, and high dielectric constant. The latter property permits construction of high-capacity capacitors of relatively small size.

The largest single use of block mica is in electronic tubes. Mica punchings having accurately positioned holes and slots, space and insulate the electrical elements therein. Other pieces center the assembly in the tube. Mica in tube applications withstands operating temperature without weakening electrically or mechanically.

Mica-insulated spark plugs are used in many aircraft engines. Block mica has many other small-volume uses where one or more of its properties make it invaluable. It is used in boiler-gage glasses because of its flexibility, transparency, and chemical inertness and for furnace inspection windows where transparency and heat resistance are needed. Other uses include domestic and industrial heating appliances and optical instruments.

Mica Paper

In recent years a new type of mica product classed as mica paper or reconstituted mica has become available. These materials are made by pulverizing scrap mica by force, heat and/or chemical action and then recombining the small particles either with or without a resin binder to form continuous flexible sheets. By the addition of resins and reinforcing backings, plate micas in tapes and wrappers have been made.

Synthetic Mica

The best electrical grade of natural mica is muscovite, especially Indian ruby mica. It is available to relatively large sheets, splits readily, and has good dielectric properties. It would therefore be expected that the mica synthesized in largest quantity would be muscovite. On the contrary, almost all mica that has been synthesized has been phlogopite. Muscovite mica has been produced in the laboratory through solid-state reaction. The largest crystals thus made, however, are of microscopic size.

There are synthetics such as fluorine mica which has the peculiar property of bonding to itself under heat and pressure. New products have been derived from it including hot pressed mica, phosphate-bonded mica, and reconstituted mica sheet. Applications of these materials have been in electromechanical items such as vacuum-tube supports for power and hydrogen-thyratron tubes. Despite the development of synthetic products, natural mica has maintained its position of importance in industrial applications and is valuable in electrical technology.

Electrical Ceramics

A ceramic product is usually regarded as being one that has been molded from its basic constituents, which in most cases includes clay, and fired to give it a hard finish. The ceramics considered herein are: alumina ceramics, beryllia, ferrites, nitrides, other oxide ceramics, porcelains, and steatites.

Both porcelains and steatites are vitreous ceramics comprising refractory inorganic silicates, produced by kiln treatment of a precompacted, finely divided mixture of naturally occurring minerals. Porcelain and steatite can cover a variety of materials with a wide range of properties, but they are fundamentally aluminum silicates and magnesium silicates respectively.

Alumina Ceramics

Alumina ceramics contain at least 85% aluminum oxide or alumina. These types have several advantages over other ceramics:

1. They are easily metallized, enabling vacuum tight joints to be made to metal.
2. They are economically fabricated into complicated shapes and by machining can be made to tight tolerances.
3. They are stable in air and oxygen over a wide range of temperatures.

Alumina ceramics have ten times the strength in compression than they have in tension. The low electrical loss characteristics of the purer aluminas give them many applications in the electronics industry. Other uses include vacuum tube envelopes and rocket radomes. The high resistivity gives good insulating properties for coil formers, resistor cores, lamp filament supports, and spark plug insulators. Metallized alumina ceramics having high surface temperatures are used as supports for cathode ray tube elements and as gun insulators in electron microscopes.

Beryllia Ceramics

Beryllia ceramics combine properties between those of some metals and ceramics, including:

ability to be joined to metals by brazing
low density and high strength
high dielectric strength
high electrical resistivity
high melting point

low neutron capture cross-section
low power factor
high thermal conductivity

Beryllia has an unusual thermal conductivity for a ceramic in that it is only surpassed by three metals—silver, copper, and gold. Beryllia is used in high temperature refractory ware, microelectronics, and in the nuclear energy industry as a moderator and reflector for power reactors. Both manufacturing cost and raw material cost of beryllia are high, the former partly due to safety precautions necessary against the toxic effect of its dust.

Ferrites

Ferrites are made under controlled firing conditions from iron oxide and the oxides, hydroxides, or carbonates of divalent metals such as cobalt, copper, magnesium, manganese, nickel, and zinc. They produce magnets which are nonmetallic and also nonconductive. They are used as ferrite rods, high frequency transformers, and as radio aerials.

Nitrides

Two of the most common nitride ceramics are boron nitride and silicon nitride. Boron nitride is an easily machined ceramic with outstanding electrical properties and is used mainly as a dielectric. Silicon nitride is coming to be of great engineering significance because of its very high strength and oxidation resistance at elevated temperature.

Other Oxide Ceramics

Magnesia A dense, sintered ceramic. It offers resistance to attack by bases. It is used in metallurgical applications. In crushed form, its high resistivity at high temperature makes it ideal as the insulator in electrical hot plates.

Titania Titania surfaces are easily polished to give a low coefficient of friction and the material is used for textile guides. A second firing at 900 °C in hydrogen transforms titania into an electrical conductor.

Titania-based Electroceramics Synthesized from barium carbonate, titanium dioxide, cobalt oxide, and lead zirconate. Titanate ceramics are used in high frequency applications. Barium titanate has a high dielectric constant, has piezoelectric and ferroelectric properties, and is used in electronic components and accelerometers.

Porcelains

There are six major porcelain types: electrical (hard) porcelains, chemical porcelains, mullite porcelains, zircon porcelains, lithium porcelains, and barium porcelains. A typical formulation for an electrical porcelain is 50% kaolin (china clay), 25% potash feldspar, and 25% quartz (milled sand). This material can be readily formed into complex shapes and large products. It exhibits high resistivity and dielectric strength and has a large glass content which makes it brittle and gives poor thermal shock resistance. Electrical porcelain applications include line insulators and switchgear for high voltages and low frequencies, suspension insulators, pin insulators, bushings, switch parts, lead-through insulators, and aerial insulators.

Silica is replaced partially by other refractory oxides in chemical porcelains. Commonly mullite, alumina, or sometimes zirconia are used as strong crystalline fillers. Properties of chemical porcelains include resistance (except to alkalis and hydrofluoric acid), low iron content, high strength, and good thermal shock resistance. Chemical porcelains are used in piping, valve, and pump parts. Mullite procelain is used in laboratory ware, burner tips, spark plugs, fuse tubes, and resistor cores. For special applications requiring dimensional stability over temperature ranges, lithium porcelains are used.

Steatites

Steatite ceramics are superior to porcelains in many respects. They are usually stronger, have better electrical properties, and can be readily formed. Steatites are the most widely used ceramics in the electrical industries and find many mechanical applications. For high frequency applications, where dielectric loss is important, steatite ceramics fluxed with alkaline earths are manufactured. Some ceramics related to porcelains and steatites find application in fuse cores, igniter tubes, resistor cores, rheostat blocks, and stable inductors.

Production Methods

A number of processes can be used to produce ceramics in various shapes. These include dry pressing, extruding, injection molding, isostatic pressing, slip casting, and throwing or jollying. Choice of method is determined usually by the size, shape, and number of components that are to be produced.

Firing of ceramics is usually in automatically controlled or oil fired kilns. The components are fed in at one end, heated to temperatures between 1100 and 1750 °C and emerge at the other end ready to handle. Linear shrinkage due to firing is about 5 − 10% and tolerances can be maintained to about ±2%. Glazing is often employed using materials such as alumina, boron or lead oxides, and silica, some of which are mixed and premelted before being ground, applied, and fired.

Refractory Ceramics

Refractory ceramic products are made from inorganic, nonmetallic materials

78

and usually are processed at high temperature during their manufacture. These materials include:

abrasive/cutting tools
all types of glass products
many types of electrical insulation
structural products

Raw materials used in the manufacture of these products are usually oxides of aluminum, magnesium, and silicon, together with silicates, and other complex oxides such as clay and talc. However, use of materials which are not oxides is increasing and includes carbides, nitrides, and silicides. There are only three types of ceramics in terms of microstructure:

glass
mixtures of crystals and glass
crystalline bodies that contain no glass

Until recently there were few ceramic products of the last type. However, with advancements in fabrication, totally crystalline ceramics are becoming more prevalent.

The major useful engineering properties of ceramics are as follows: hardness, refractoriness, chemical inertness, electrical, nuclear and optical properties, and high strength both at room and elevated temperatures. Ceramics usually have poor impact resistance and tensile strengths are much lower than compressive strengths, thus design should favor compression load applications.

Hardness

Ceramic products in general are considerably harder than most materials. Hardness makes certain ceramics extremely useful as abrasives, cutting tools, and for producing surfaces which must withstand severe abrasive action. Boron nitride, a recently developed material, is nearly as hard as diamond.

Refractoriness

Ceramics are the materials having the highest known melting points. Therefore such materials find use in furnaces and enclosures for containing high temperatures. Although most ceramics have high melting points, a few do not. Some glasses melt below 540 °C and there are several glasses which are liquid at room temperature. Hafnium carbide, a ceramic, has the highest melting point (3900 °C) of all known materials.

Chemical Resistance

Most ceramics are highly resistant to all chemicals except hydrofluoric acid and, to some extent, hot caustic solutions. Organic solvents do not affect them.

Strength

Most ceramics have high compressive strengths. Strengths vary greatly with composition and porosity. Elevated temperature strength retention makes certain ceramics desirable for use where high thermal stresses are encountered. Titanium diboride has been reported to maintain a strength as high as 240 MPa up to 2000 °C, the strongest material known at this temperature.

Impact Resistance

In general, ceramic materials exhibit low impact resistance. However, with proper design they are used in certain applications where impact conditions are present and other environmental conditions require the use of ceramics.

Electrical Properties

Ceramics are available which include low loss, high frequency insulators, conductors, semiconductors, ferroelectrics, and ferromagnetics. Although only several decades ago electrical ceramics were restricted to insulators, current applications include thermistors, thermoelectrics, transistors, piezoelectric transducers, and storage cells in memory systems.

Nuclear Properties

Ceramics have significant use in nuclear applications since they are refractory, chemically resistant, and because different compositions offer a wide range of neutron capture and scatter characteristics. They are used in fuel elements, controls, moderators, and shielding.

Optical Properties

Glasses have been historically the major material used for windows and optical lenses. Special glasses, however, have been developed for selective transmission or absorption of particular wavelengths, such as infrared and ultraviolet. Fused silica windows and windows made from large synthetic crystals of alumina are used for high temperature applications. Photochromic glasses containing silver halide are used for spectacle lenses which automatically darken in sunlight and then become clear indoors.

Cermets

Cermets are a class of material containing both ceramic and metal, and hence combine properties of the two. The ceramics usually include oxides, carbides, and nitrides; the metal may be either pure or an alloy. Further, cermets may contain one or more of either or both types of constituents and the composition may vary from predominantly ceramic to predominantly metallic.

Development of cermets was undertaken to provide materials for service at higher temperatures than those which could be withstood by nickel and cobalt base alloys. The principle employed was that of combining the ability for high temperature strength retention of ceramics with the ductility and toughness of metals. Although resistance to impact damage of cermets is inferior to that of high temperature alloys, they are materials with unique properties and large field of utility. Notwithstanding that much work has been performed on many material combinations, only a few have major applications. These include titanium carbide base cermets, alumina base cermets, and uranium dioxide base cermets developed for nuclear reactors.

Properties

Titanium carbide base cermets feature low density, high modulus of elasticity, and high strength at high temperature. These properties, high hardness, and excellent resistance to oxidation are important in applications such as high speed metal cutting and finishing. The stress – rupture properties of titanium carbide base cermets are outstanding in comparison to those of nickel and cobalt base high temperature alloys. In rotating parts such as turbine buckets and impellers, material density contributes directly to stresses. In this respect, these cermets have a particular advantage.

Alumina base cermets have slightly better oxidation resistance than the titanium carbide base cermets; further, they are highly corrosion resistant. Chromium carbide base cermets have very high resistance to chemical corrosion in many media.

Cermets require powder metallurgy techniques for fabrication, thus limiting size and complexity of components. The high melting points of the ceramic constituents make melting impractical. Powder metallurgy methods provide the means of retaining fine-grained uniform structure without decomposing the ceramic phase. Fabrication methods used include cold press and sinter, cold press and sinter followed by infiltration, hot press, extrusion, and slip casting. Auxiliary to several of these methods are techiques such as turning and drilling. Secondary operations are usually performed in the presintered condition.

Applications

Aside from high temperature applications in turbine buckets, nozzle vanes, and impellers for turbines, cermets are used in temperature sensing elements and thermostats where oxidation resistance, low coefficient of thermal expansion, and ability to be welded directly to alloys are important. Titanium carbide base cermets are also used for bearings and in liquid metal pumps, hot flash trimming and hot spinning tools, and hot rod mill guides. They are also used in some cryogenic applications such as for rotary seals. Alumina base cermets are used as thermocouple protection tubes for molten metal applications.

6
Composite Materials

'Composites' are man-made material systems consisting of a reinforcement and a surrounding binder or matrix fabricated into a useful end product. Much the same as familiar examples of plywoods and concretes, these material are nonhomogeneous but even more highly anisotropic, i.e., properties are strongly dependent on direction. Further, the advanced or 'high performance' composites are comprised of high strength, high stiffness, or special purpose fibers held together by strong, tough polymers or plastics, ceramics or metals.

These materials are all relatively new; however, their principle goes well back in man's history. The early Egyptians made strong, stiff, light boats from bonded plywoods using animal glues. Ancient China had the first laminated archery bows. The Damascene blade was laminated of various sheets of metal hammered into a strong and resilient composite superior to that of an isotropic metal sword. There are many other examples, some dating back 2000 years.

What is new, however, in the materials field is the explosive growth of fiber reinforced plastics since the early 1940s. This growth was vigorously stimulated by wartime needs. The very early glass fiber cloth/polyester resin composites have only within the past decade or so given birth to many other fibers and a host of superior plastics including structural epoxies and cryogenic and high temperature resin systems. Boron fiber, mechanically a very strong material, was developed during this period by accident from studies on chemical additives for solid propellants. Graphite fiber, in a number of respects superior to boron, has only been available for little more than the past decade.

Types of Composite Material

Composites may be classified according to the following scheme.

1. Reinforcement geometry
   - (a) Particulate fillers
   - (b) Flakes
   - (c) Whiskers (short length to diameter (L/D) ratios)
   - (d) Staple fibers (intermediate L/D ratios)
   - (e) Continuous fibers (infinite L/D ratios)
2. Reinforcement material
   - (a) Asbestos

82

    (b) Beryllium wire
    (c) Glasses
    (d) Sapphire
    (e) Silicon carbide
    (f) Stainless wire
    (g) Tungsten wire
    (h) Boron   ⎰ advanced composites
    (i) Graphite ⎱
  3. Matrix material
    (a) Plastics
    (b) Ceramics
    (c) Metals

The advantages of composites over conventional metallic alloys are as follows:

    increased strength and stiffness
    lower densities
    high fatigue resistance
    high internal damping
    relative ease of fabrication
    capability for 'design tailoring', i.e., putting material strength and stiff-
       ness where it is needed in the design
    good corrosion resistance
    good electrical and thermal resistance

Principal disadvantages of most composites are their costs and that sophisticated design procedures are required to cope with anisotropy. There are other problems associated with brittleness and attachment methods as there are with quality control and inspection procedures.

Applications

Numerous aircraft and aerospace applications of advanced composites have been made since 1960 – 65 and have included, for instance: buckling-critical fuselage structure, strength-critical aircraft tail structure, wing flaps, and control surfaces, and filament would high pressure rocket motor cases and positive expulsion tanks. Reinforcements for local penetrations and cutouts have received much attention. Hybrid composite/metal structures have also been produced. Numerous other applications of composites include compressor blades, helicopter and space structures, air cushion vehicle and marine structures, and rocket nozzles.

Composites have had a profound effect on aviation and space efforts and are affecting many other fields of engineering. Structures, motors, ships, automobiles, bridges, railroads, building construction, sporting goods, medical prosthetics, etc., are likely to benefit from stronger, lighter, and cheaper materials. It is but a matter of time until the application 'market volume' develops to the point where composites becomes much cheaper than at present. Their development and exploitation should continue for many years to come.

References

DIETZ, A.G.H. *Composite Engineering Laminates*, MIT Press, Cambridge, Mass., 1969

ASHTON, J.E., HALPIN, J.C. and PETIT, P.H., *Primer on Composite Materials: Analysis*, Technomic, Stamford, Conn., 1969

ASHTON, J.E. and WHITNEY, J.M. *Theory of Laminated Plates*, Technomic, Stamford, Conn., 1970

AMBARTSUMYAN, S.A. *Theory of Anisotropic Plates*, Technomic, Stamford, Conn., 1970

WENDT, F.W., LIEBOWITZ, H., and PERRONE, N., *Mechanics of Composite Materials*, Symposium on Naval Structural Mechanics, Pergamon Press, New York, 1970

TSAI, S.W., HALPIN, J.C. and PAGANO, N.J., *Composite Materials Workshop*, Technomic, Stamford, Conn., 1968

JONES, R.M. *Mechanics of Composite Materials*, McGraw-Hill, New York, 1975

VINSON, J.R. *Composite Materials and Their Uses in Structures*, Applied Science Publishers, London, 1975

SALKIND, M.J. Applications of composite materials, ASTM STP 542, 1973

SALKIND, M.J. Composite materials: testing and design, ASTM STP 460, 1969

LUBIN, G. *Handbook of Fiberglass and Advanced Plastic Composites*, Van Nostrand, Princeton, N.J., 1969

BROUTMAN, L.J. *Modern Composite Materials*, Addison-Wesley, Reading, Mass., 1967

DAVIS, L.W. and BRADSTREET, S.W. *Metal and Ceramic Matrix Composites*, Cahners, Boston, Mass., 1970

See also UK and US journals of composite materials and publications of the American Institute of Aeronautics and Astronautics, the American Society of Mechanical Engineers, the American Society for Testing and Materials, and the Society of the Plastics Industries.

7
Plastics

The period following the Second World War may well be termed the 'Plastics Age' since plastics have become so important in this time. It is expected that their volume usage will exceed that of metals in the mid-1980s; however, because of dwindling world petrochemical supplies, their future in the twenty-first century appears uncertain.

Organic in nature, plastics are usually made from large chain-like molecules built up from molecules of carbon, hydrogen, oxygen, and nitrogen. Properties of plastics depend greatly on the molecule size and the arrangement of atoms within the molecule. Other types of plastics—e.g., the amino resin types: melamine and urea—have large three-dimensional molecular structures. Consequently, a very wide range of plastics with differing properties is available and these plastics find many uses. Typical properties are low density and low cost, moderate strength, corrosion resistance, heat and chemical resistance, good impact strength, insulating properties and formability.

Plastics may be grouped into three categories:

1. *Thermoplastics*. Solid at room temperature, these materials soften with heating and eventually become liquid. Fluidity is utilized in processing finished parts.
2. *Thermosetting resins*. These materials are of little use in their natural state, but they are made useful by heating and curing, when they become stiff and hard. Thermosets remain stable and the process is irreversible.
3. *Elastomers*. These materials withstand repeated elongation, of up to 200%, yet may show complete strain recovery upon release of applied stress.

Major plastics-producing countries are Japan, the USA, the Federal Republic of Germany, Great Britain, and Italy. A guide to the overall characteristics of plastics is given in Table 9. This chapter describes plastics in the order as given in Table 9, with their properties and applications.

Acrylonitrile Butadiene Styrene

Acrylonitrile butadiene styrene (ABS) plastics are a family of opaque thermoplastics offering a balance of properties, the most outstanding being impact resistance, tensile strength, and scratch resistance. Two methods of manufacture are employed for ABS plastics:

Table 9   Guide to Selection of Plastics

Plastics	High impact strength	High scratch resistance	Good weatherability	Readily electroplated	High tensile strength	Resists acids	Resist alkalis and solvents	Good colorability	Low friction
Acryolonitrile butadiene styrene	√	√	√	√				√	
Acetals	√				√		√	√	√
Acrylics						√	√	√	
Alkyds									
Cellulosics									
Cellulose acetate								√	
Cellulose acetate butyrate			√						
Cellulose acetate propionate	√								
Cellulose nitrate			√					√	
Ethyl cellulose	√								
Diallyl phthalates						√	√		
Epoxies		√			√	√			
Fluorocarbons									√
Melamines		√						√	
Nylons									√
Phenolics						√			
Polycarbonates								√	
Polyesters	√		√			√			
Polyethylene									
Polyimide					√	√			√
Polypropylene						√	√	√	
Polystyrene						√	√	√	
Polysulfone	√				√				
Polyvinyls									
Polyvinyl acetate									
Polyvinyl butyral									
Polyvinyl chloride									
Vinyl copolymers		√							
Silicones						√			√
Ureas		√				√	√		

86

Plastics	Excellent optical properties	Low moisture absorption	Good adhesives	Good electrical properties	Arc and track resistant	Good corrosion resistances	High temperature capability	Low density	Large volume, low cost usage
Acryolonitrile butadiene styrene									
Acetals									
Acrylics	√	√	√						
Alkyds		√		√	√				
Cellulosics									
Cellulose acetate	√								
Cellulose acetate butyrate						√			
Cellulose acetate propionate	√								
Cellulose nitrate									
Ethyl cellulose		√							
Diallyl phthalates			√	√					
Epoxies				√		√			
Fluorocarbons				√		√	√		
Melamines		√	√	√	√				
Nylons			√				√		
Phenolics		√	√	√	√	√	√		
Polycarbonates						√	√		
Polyesters		√		√		√			
Polyethylene									√
Polyimide			√				√		
Polypropylene							√		
Polystyrene		√				√		√	
Polysulfone				√					
Polyvinyls									
Polyvinyl acetate			√						
Polyvinyl butyral	√	√	√						
Polyvinyl chloride									√
Vinyl copolymers									
Silicones				√	√	√	√	√	
Ureas			√						

1. *Copolymerization.* A butadiene/acrylonitrile latex is added to a styrene/acrylonitrile latex and is coagulated and spray dried to give a copolymer.
2. *Graft polymerization.* Styrene and acrylonitrile are added to a polybutadiene emulsion and the mixture is stirred and warmed. A water soluble initiator is then added and the mixture polymerized. The resulting 'graft polymer' has a higher impact resistance than the copolymer, but has lower rigidity and hardness.

Control of the processes facilitates tailoring ABS to suit requirements.

Properties

The ABS plastics are ductile and are characterized by high impact strength, moderate tensile and compressive strengths, marked dimensional stability, and extremely good corrosion and chemical resistance. They are very rigid and retain their properties over a wide range of temperatures, frequently as low as -40 °C for refrigerator parts.

ABS is easily electroplated. Flow characteristics give good moldability and colors available are unlimited with high gloss stain resistant finishs. Weatherability is outstanding—most environments have little or no effect on ABS, although prolonged strong sunshine may cause embrittlement. Resistance to creep and cold flow is moderate to good. Acoustic damping characteristics are excellent. Flammability is classified as slow burning, but ABS may be made flame resistant by blending with PVC.

Applications

ABS is an extremely versatile material. It is suitable for blow and injection molding. It may also be extruded, vacuum formed, cold formed, or stamped. Casting is not used. ABS is frequently used in pipes and pipe fittings because of corrosion resistance and has vast domestic applications, such as refrigerator linings and fittings, vacuum cleaners, telephones, etc. Its impact resistance and dimensional stability make it suitable for automobile and motorcycle parts, e.g., instrument panels, safety helmets, and panels. ABS is used in selective electroplating of certain parts such as grills, and door handles, eliminating paint applied over chrome. Other recent developments include self-extinguishing grades, optically clear grades and polycarbonate alloys.

Acetals

Acetals are a relatively new thermoplastic, first becoming available in 1960. In appearance they are similar to nylon, with which they compete in applications such as gears, bearing, hardware components, and business machine assemblies. The resin is a linear polymer of formaldehyde that is highly crystalline. Acetals may exist as a homopolymer (e.g., Delrin) or as a copolymer (e.g., Celcon).

Properties

Acetals have excellent mechanical properties which enable them to be used in place of metals in many applications. They have a high tensile strength which is retained over at least several years in air/water exposures at temperatures of 95 °C. They have high impact resistance which is only slightly affected by subzero temperatures. Their stiffness is high, and creep and fatigue resistance exceed that of any other thermoplastic. The coefficient of friction is low and abrasion resistance is comparable to metals.

Acetals resist solvents and most alkalis but are attacked by acids. Colorability is good and discoloration by industrial oils and grease is not a problem. Moisture absorption is low, weatherability is good, but short-term exposure to ultraviolet radiation causes chalking of the surface, while prolonged exposure affects strength.

Applications

The above properties of acetals make their use possible in place of metals in many applications. Some typical uses are as gears, bearings, cams, switches in automotive and business machines, hardward components and screws. Lubricity may be improved for special applications by the addition of fluorocarbons (e.g., for bearings), and stiffness and dimensional stability are improved by incorporating glass fibers (e.g., for business machine assemblies and housings).

Normally acetals are injection molded or extruded, but parts are sometimes machined from stock shapes. In this aspect, acetals resemble free-cutting brass —they may be sawed, drilled, turned, milled, shaped, reamed, etc., without cutting fluid. They are tough enough to withstand riveting and staking of metal fasteners and may be joined by spin-welding, by ultrasonic welding, or by bonding with epoxy resins.

Acrylics

Acrylics are thermoplastics which offer excellent optical characteristics, resistance to environmental conditions, and ease of forming. They are manufactured from methyl methacrylate monomers found in the by-products of petroleum, agricultural, and synthetics industries, and are copolymerized for form acrylics such as polymethyl methacrylate, known by trade names such as Lucite, Perspex, and Plexiglass.

Properties

Acrylics transmit about 92% of light and have a high refractive index. They offer very good dimensional stability and are serviceable to 110 °C, however, they show a large thermal expansion—about seven times that of steel.

Resistance to weathering is excellent but high intensity ultraviolet light exposure produces crazing, which may be alleviated by annealing. High tensile strength and good impact resistance enhance its usefulness.

Arcylics are unaffected by alkalis, industrial oils, inorganic solvents, and weak acids or alcohol, but concentration of acid or alcohol results in deterioration. Moisture absorption is low.

Applications

Acrylic sheet is used for signs and displays, light fittings (particularly street lighting), dome and sky lights, instrument panels, and safety shields. Recently it has been applied to solar absorption systems in place of glass because of its greater clarity. However, the specific acrylic must have a resistance to sunlight greater than normal and thermal expansion must be taken into account in assembly.

Acrylic powder may be molded for automobile trims such as tail light lenses, instrument panels and dials, etc. High impact powders are used for piano ivories and business machine components. Acrylic resins are used in paint formulations and adhesives. In recent years, acrylics have been used in architecture, paintings, and sculpture.

Acrylics are available in the following forms:

1. cast sheet which may be heated and formed, sawn, drilled, and machined —it is strong, stable, and transparent
2. molding powder, which may be injection, extrusion, or compression molded
3. high impact molding powders

Alkyds

Alkyds are thermoset molding materials. They are made by combining a monomer with unsaturated polyester resins and additives or fillers.

Properties

Like most thermosets, alkyds are hard and stiff and retain their mechanical and electrical properties at elevated temperatures. Their properties depend greatly on the fillers and manufacturing processes used. Typically, tensile strength is low, whereas compressive strength is much higher, by a factor of 4 – 5. Outstanding properties are their arc and track resistance, low moisture absorption, and retention of electrical properties when wet. Only small changes in dielectric loss factor are caused by temperature, Dimensional stability is very good.

Applications

There are three basic types of alkyd classified according to raw materials, each of which has its applications.

Granular types Mineral fillers are used to obtain these materials, which have very good dielectric properties, heat resistance, excellent dimensional stability, and insulating properties under severe conditions. However, strength is low. Their main uses are in high-grade electrical insulation. Granular alkyds have been used as molded rotors, distributor caps, and coil tops in automotive electrics. Granular types are used mostly in compression molding, with some grades being transfer molded. Self-extinguishing grades are available.

Putty types Available as soft putty-like sheets, these types are easy to mold with no distortion and they require very low pressures. Their major use is for molded encapsulation of small electronic parts and for intricate sections. In such uses they provide electrical insulation and moisture protection.

Glass fiber and reinforced types These types are made from rope and bulk raw materials. The reinforcement enhances strength, arc, impact and heat resistance, but restricts moldability. Rope alkyd is a putty-like material containing reinforcement with glass or synthetic fibers and offers good mechanical and electrical properties while retaining good molding characteristics. Bulk alkyd has a much higher concentration of reinforcement, making it suitable for very high impact applications but at the expense of difficult molding. Generally compression and plunger molding are used. Bulk types are used for circuit breaker housings and switchgear installations. A humidity resistant grade has also been developed recently.

Alkyds are available in a broad range of colors, but decorative applications are few because of cost and mediocre surface finish. Alkyds resist weak acids and organic solvents but do not resist alkalis or concentrates. Bonding of alkyds is performed only with epoxy resins.

References

BECK, R., *Plastic Product Design*, Van Nostrand Reinhold, New York, 1970
BRISTON, J.H., *Plastic Films*, Iliffe, London, 1974
MILBY, R. V., *Plastics Technology*, McGraw-Hill, New York, 1973

Cellulosics

Cellulosics are thermoplastics and include cellulose acetate, cellulose acetate butyrate, cellulose acetate propionate, cellulose nitrate, and ethyl cellulose.

Cellulose Acetate (CA)

CA is an amber-colored, transparent material first available in the 1930s. It is formed by the reaction of cellulose, acetic acid, acetic anhydride, and a catalyst. The end product is flake acetate which can be formed into fibers and films.

Additives are required for various applications, e.g., plasticizers for toughness and heat resistance. Colorants and stabilizers may also be added. CA continuous fibers or 'acetate' form the second most important fiber to rayon.

Characteristics CA is available in translucent and transparent forms and in opaque colors having mottled or pearly effects. Depending on plasticizer content, it will absorb moisture. CA is non-toxic. It is fast drying at ambient temperature and in lacquers provides ease of application and good adhesion. CA film is heat sensitive.

Applications Extruded and cast film and sheet are used for protective packaging, e.g., containers, sealing bottles, etc., because of transparency, non-toxicity, non-flammability, and ease of fabrication. Extruded rod is used for tool handles and machine parts. Extruded profiles are used in trim and wear strips. Lacquers are used for wallpaper, textiles, and floor and linoleum finishes. Industrial goggles, machine safety guards, and photographic film are other uses. Because of low pressure drop and high absorption, CA filaments are used in cigarette filter tips.

Other features of CA are that it is less hygroscopic than wool, viscose, rayon, and silk, that special color effects can be obtained with dyes, and that it is supple and soft feeling to the touch. Other processing methods employed with CA are blow molding, foaming, injection molding, rotational molding, and thermoforming.

Cellulose Acetate Butyrate (CAB)

CAB is formed similarly to CA but with butyric and acetic anhydrides. Properties are varied by altering proportions of acids and anhydrides and the degree of hydrolysis used. The butyrate is produced by adding plasticizer in a hot mixing process.

Characteristics CAB resembles CA in appearance but is less temperature sensitive and more resistant to solvent attack. It has higher impact strength, lower moisture absorption, and is slower burning than CA. CAB is formed by continuous extrusion and is available in granular form for injection and compression molding. It can be colored to any degree.

Applications A major use for CAB is as extruded pipe for natural gas lines and electrical cable lines because of high corrosion resistance and low friction factor. Sheets are used for sign production, street light globes, and outdoor signs requiring weather resistance. Transparent CAB is used in television screens, tool handles, and lenses for instrument panel lights. It is molded and extruded into many products such as blister packaging, pens, and containers.

CAB is tough, decorative, protective, smooth to the touch, and readily colored. The odor developed from butyric acid in enclosed spaces can be obnoxious; hence most CAB is used outside.

92

Cellulose Acetate Propionate (CAP)

CAP is formed similarly to CA but with acetic anhydride and propionic anhydride. CAP is recovered after hydrolysis.

Characteristics CAP has good dimensional stability, moldability, hardness, and toughness. It has excellent impact and flexural strength, good shock resistance, and low moisture absorption. Dimensional changes due to temperature and humidity are small. Creep is negligible below critical stresses and temperatures, hence continuous use under high load or very high temperatures is not recommended. CAP is slow burning and gives excellent service in outdoor uses if appropriate grades are used.

Applications Injection molded CAP is used for small articles, e.g., telephones, steering wheels, knobs, toothbrush handles, pen barrels and caps, fuel filter bowls, flashbulb shields, and tough transparent machine components. Other products are safety goggles and toys. Plastic rod is used to make handles for screwdrivers, wrenches, and machine controls. Thin walled tubing is used in packaging and transparent extruded sheet is used in heavy duty blister packaging.

Cellulose Nitrate (CN)

CN was the first successful plastic and dates back to 1869, the beginning of modern plastics. Commercial rayon was produced in 1884 from the textile fiber but was short lived because of flammability. CN lacquers for the automobile industry were developed in the 1920s. CN is prepared by converting alcohol groups of cellulose into nitrate esters. This requires the presence of nitric acid and sulfuric acid for hydration and to suppress ionization of the nitric acid. Three stages are involved in its manufacture: pretreatment of cotton, nitration, and aftertreatment and dehydration. Factors which affect these processes are: strength of mixed acids, temperature, time, and moisture content.

Characteristics CN products are inexpensive, easily worked, readily machined, cemented, swaged, blown, and formed. They cannot be injection molded. All colors are available as well as transparent, translucent, mottled and opaque forms. Conventional machining processes are applicable provided cooling water is used, e.g., cutting, sawing, punching, drilling, drawing, turning, printing, embossing, and polishing. CN is unaffected by hydrocarbons, animal, vegetable, and mineral oils, and low concentrations of mineral acids at normal temperatures. CN turns yellow or brown on exposure to sunlight and becomes brittle due to loss of plasticizers. It burns rapidly and completely once ignited, is decomposed by strong alkalis and acids, and is dissolved or softened by many alcohols, ketones, and esters. CN lacquers are fast drying at ambient temperature, have good outdoor durability, and are easy to apply. It is manufactured in sheets and rods which are tough and serviceable.

Applications Applications of CN include guncotton (an outdated explosive), smokeless powders, and soluble pyroxylin—used in common lacquer solutions, e.g., car painting. When plasticized with camphor, CN forms the thermoplastic xylonite known as celluloid which is used for combs, toothbrushes, table tennis balls, plastic playing cards, spectacle frames, etc.

Ethyl Cellulose (EC)

Ethyl cellulose is a cellulose ether in which ethyl groups replace hydrogen in the hydroxyl groups of glucose residues. It is formed by treating alkali cellulose with ethyl chloride or ethyl sulfate. EC is prepared by treatment of purified wood cellulose or cotton having high cellulose content. Flakes thus formed are purified and blended with plasticizers, stabilizers, and pigments to obtain various molding and extrusion formulations.

Characteristics EC possesses toughness, thermoplasticity, and can withstand shock. It has a low flammability, good electrical and dielectric properties, good water resistance, and is compatible with a variety of natural resins and mineral oils. EC is inert to all alkalis and may be used over a range of temperatures and humidity conditions. It retains high flexibility at low temperatures, which makes it useful for certain gaskets and fabric materials. EC is available in an almost unlimited range of colors, as with the other cellulosics. Molding and extrusion processes are used.

Applications EC is used in lacquers, varnishes, and adhesives to improve toughness and flexibility. Hair sprays are an example. It is added to waxes to give increased strength and raise the melting point. Cable and paper coatings and frozen food cartons are other uses. The toughness and resistance of EC to nitroglycerine makes it suitable for military uses. Helmets, gears, rocket components, rollers, slides, sweepers, and handles are further uses.

Diallyl Phthalates (DAP)

Diallyl phthalates (sometimes called allylics) are thermosetting materials. DAP resin is the first all-allylic polymer commercially available as a dry, free-flowing white powder. Allylic resins enjoy certain advantages over other plastics which make them useful in various applications. For example, under severe temperature and humidity conditions, they exhibit superior electrical properties even after repeated exposure. They also exhibit excellent post-mold stability and low moisture absorption while showing good resistance to solvents, acids, and alkalis.

Properties depend on the type of filler used. For example the highest heat resistance and tensile strengths are obtained by using asbestos or glass fiber fillers. Diallyl phthalates are used in molding and industrial and decorative laminates because of the combination of low shrinkage and superior electrical and physical properties. Molding compounds of DAP resin and various fibers such as asbestos, Orlon, Dacron, and glass are used for resistors, connectors, potentiometers, switches, and insulators. Arc and track resistance are features of such applications. In industrial laminates, DAP resins with woven and non-woven fabrics of glass and synthetic fibers are used in tubing, ducting, junction boxes, and aircraft and missile parts. Decorative laminates can be made from glass cloth or other woven and nonwoven materials. DAP resins give a permanent finish to high grade wood veneers.

Epoxies

An epoxy resin is defined as any molecule containing more than one α-epoxy group and capable of being converted to a useful thermoset form. The term is used to indicate the resins in both the thermoplastic (uncured) and thermoset (cured) state. Cured epoxies are a major structural type in plastics technology.

Properties

A number of properties have led to the rapid growth in the use of epoxy resins and their use in many industries.

Liquid Resins These are low-viscosity liquids which readily convert to the thermoset phase upon mixture with a curing agent. There are other liquid resins—acrylics, phenolics, polyesters, etc.—which cure in a similar fashion, but epoxy resins possess a unique combination of properties.

1. *Low viscosity.* The liquid resins and their curing agents form low-viscosity, easy-to-process systems.
2. *Easy cure.* Epoxy resins cure quickly and easily at practically any temperature from 5 to 150 °C depending on selection of curing agent.
3. *Low shrinkage.* One of the most important and advantageous properties of epoxy resins is their low shrinkage during cure. Epoxy resins react with very little rearrangement and with no volatile by-products.
4. *High adhesive strengths.* Epoxy resins are excellent adhesives. The best adhesive strengths in contemporary plastics technology are obtained without the need for long curing times or high pressures.
5. *High mechanical properties.* The strength of properly formulated epoxy resins usually surpasses that of other casting resins. This is in part due to their low shrinkage, which minimizes curing stresses.
6. *High electrical insulation.* Epoxy resins are excellent electrical insulators.
7. *Good chemical resistance.* The chemical resistance of cured epoxy resin depends considerably on the curing agent used. Outstanding chemical resistance can be obtained by proper specification of the material.
8. *Versatility.* Epoxy resins are highly versatile. The basic properties may be modified in many ways: by the blending of resin types, by the selection of curing agents, and by the use of modifiers and fillers.

Solid Resins The chief use of solid epoxy resins is in solution coatings. The high molecular weight materials are cooked with conventional drying oils or reacted with other resins resulting in toughness, scuff resistance, and chemical resistance. Room temperature curing films have been produced which provide properties equal to or exceeding those of many baked-type finishes.

The excellent adhesion of epoxy resins, ease of cure, mechanical strength, and high chemical resistance are advantages of the solid and liquid resins.

Applications

Because of their versatility, epoxies are used in numerous applications such as:

> Adhesives for aircraft honeycomb structures, for paintbrush bristles, and for concrete topping compounds
> Body solders and caulking compounds for repair of plastic and metal boats, automobiles, etc.
> Casting compounds for fabrication of short-run and prototype molds, patterns, and tooling
> Caulking and sealant compounds in building and highway construction applications and where high chemical resistance is required
> Potting and encapsulation compounds, impregnating resins, and varnishes for electrical and electronic equipment
> Laminating resins for airframe and missile applications, for filament-wound structures, and for tooling fixtures

Epoxy-based solution coatings are used as maintenance and product finishes, marine finishes, masonary finishes, structural steel coatings, aircraft finishes, automotive primers, can and drum linings, and in many other industrial applications.

Fluorocarbons

Fluorocarbons are compounds of carbon in which fluorine instead of hydrogen is attached to the carbon chain. The resulting compounds are very stable to heat and chemicals. Fluorocarbon plastics include polytetrafluoroethylene (PTFE or Teflon), fluorinated ethylene propylene (FEP), and polyvinylidene fluoride (PVF).

PTFE

The most important nonelastomeric fluorine-containing plastic is PTFE, a white, nontoxic, waxy solid which can be pigmented. Because of its high viscosity, PTFE cannot be processed by techniques normally used for thermoplastics such as injection molding. Instead, the techniques used consist of cold shaping operations followed by sintering, during which the polymer particles fuse and coalesce, followed by cooling.

96

PTFE is virtually immune to chemical attack by molten or strong solutions of alkalis throughout its usable temperature range. Electrical properties are exceptional. PTFE is nontracking. It has the lowest known coefficient of friction and is used as a release agent. Water absorption is less than 0.10%. The usable temperature range extends from liquid nitrogen temperatures up to 260 °C. PTFE melts at 330 °C. PTFE does not have good mechanical properties. It has low resistance to wear, deforms severely under load, and has a pronounced tendency to creep. Such problems are overcome by the use of fillers.

PTFE is used in the chemical, food, and pharmaceutical industries where its chemical inertness prevents corrosion and contamination. PTFE's low coefficient of friction allows its use in the production of bearings of the sleeve and slip-joint types and in ball and roller bearing components. Its nonstick properties are most widely exploited for cookware.

Fluorinated Ethylene Propylene (FEP)

FEP is a true thermoplastic and can be processed in conventional extrusion and injection-molding equipment. It has unusual electrical properties in that both its dielectric constant and its dissipation factor are not only extremely low, but remain constant over a wide range of temperatures and frequencies. The other properties of FEP are similar to PTFE. Its melting point is 268 °C. Typical applications include small components in electrical and electronic equipment and as primary insulation for conductors.

Polyvinylidene Fluoride (PVF)

PVF has a lower melting point than FEP but it has greater room temperature tensile, compressive, and impact strengths. It is self-extinguishing and can be formed by the same processes as FEP.

Reference

KINNEY, G.F. *Engineering Properties and Applications of Plastics*, John Wiley, New York, 1957

Melamines

The chemistry of producing melamines—thermosetting resins—is complex, although molding and laminating processes closely parallel those used with ureas and phenolics.

Properties

Melamines have good electrical properties and are used for components

97

requiring resistance to surface tracking and damp environments. They are easy to mold, more so than urea formaldehyde resins, and exhibit dimensional stability. Other characteristics include good mechanical strength, chemical resistance, arc and heat resistance, water resistance—cycling through dishwashers or steam cleaners is not harmful—and good colorability.

Applications

Melamine is an almost universal choice for impregnated paper used for decorative panels and engraved signs. Applied as a clear liquid penetrant to previously printed sheets or as a pigmented resin to plain sheets, the resin-treated overlay is hot pressed over less expensive, phenolic-impregnated core sheets. A special overlay sheet of clear-treated stock can be applied to either type of panelling to increase wear resistance.

Panels are used for tables, counter and sink tops, wall coverings, desk tops, furniture sections, and similar uses requiring highly decorative and durable finishes. Plasticized grades are furnished for applications in which curing is necessary, as in sink installations. Moderate heating makes the cured resin sufficiently elastic to permit the forming of large curves. Other uses of melamine laminates include engraving stock, laminated in sandwich form, with contrasting outer ply cut to expose the core sheet.

Besides presenting a hard, mar-resistant surface, melamine laminates are superior to other resins in offering a composition imprevious to alcohol, oils, grease, mild acids, and detergents. Industrial laminates using cotton, glass, or asbestos textiles for reinforcement are widely used for circuit boards, and as mounting panels for electrical and chemical equipment. Very large quantities of melamine powder and liquid adhesives are produced annually. Although more expensive than urea glues, melamine adhesives have better water resistance, which accounts for some usage in plywood.

Nylons

Nylon is a household word throughout the world. Nearly everyone is familiar with nylon fibers used for wearing apparel, brush bristles, and carpet. Less widely known are the commercial and industrial uses of nylon molding resins, coatings, adhesives, and films.

Polyamide is the chemical term used to describe linear polymers in which the structural units are connected by amide groups. The first polyamide produced in continuous filament form was used for hosiery and undergarments as a replacement for natural silks. Marketing of the original products was accompanied by glowing advertisements proclaiming the 'new miracle material made from coal, air and water'. Figuratively this was true, for indeed the early polyamide had its origins in these common materials.

The trade name 'Nylon' was adopted, but today the term nylon is used generally to describe any polyamide capable of forming polymers. The types of nylons are so numerous that a numbering system to describe them has been adopted. It involves digits denoting the number of carbon atoms in the parent chemicals.

Properties

General characteristics of nylons are as follows.

Transparency   The natural state of nylon molding and extrusion resins is translucent, beige, or off-white. Extruded films find applications in prepared food packaging where foods are boiled in the package.

Anti-drag and Antifriction Properties   Extruded nylons provide abrasion and cut-through resistance and they can be pulled through complicated conduit paths because of inherent anti-drag properties. Low friction, a property common to all nylons, makes them suitable for gears, rollers, cams, door latch components, and many other moving or bearing parts.

Heat Resistance   Most grades are self-extinguishing and impart an odor of burning wool. Heat resistance based on deflection under load is from 65 to 180 °C with types 6 and 6/6 exhibiting the higher values. Continuously exposed to dry heat, nylons embrittle at about 120 °C.

Moisture Absorption   All nylons absorb moisture to a degree depending on formulation, and reach an equilibrium between 0.20 and 0.25%. Type selection and design tolerances thus become critical when moving parts are required to operate in a humid environment.

Chemical Resistance   Nylons are generally resistant to many chemicals, notably gasoline, liquid ammonia, acetone, benzene, and organic acids. They are attacked, lose strength, and swell when exposed to chlorine and peroxide bleaches, nitrobenzene, and hot phenol. Nylons are not recommended for extended exposure to ultraviolet light, hot water, and alcohols. They are resistant to moth larvae, fungus, and mildew.

Melting Point   Unlike most thermoplastics, nylons are highly cyrstalline with sharply defined melting points at which they become extremely fluid and free flowing. Nylons degrade rapidly when held too long in extruder heating chambers or injection machines. Parts are discolored and lose strength and/or chemical resistance.

Applications

The 'sheer' quality obtainable from fine nylon fiber in clothing is caused by high tensile strength and elasticity imparted by cold-drawing. As filaments are drawn from the melt, they are stretched as they cool. This orients the crystalline structure, aligning the molecules along the fiber axis. Heavier filaments are used for insect screening, surgical thread, parachute fabric and line, tents, and carpet. Nylon woven fabrics coated with polyvinyl chloride are used for rainwear, aircraft covers, and chemical-resistant clothing. Used with polyesters, nylon fabrics provide lightweight laminates for aircraft canopy attachments and other applications. Nylon flake dissolved in phenolic resin improves the flexibility of laminated fishing rods.

Phenolics

Phenolic resins are among the oldest of plastics and were the first to be commercially exploited. They are used principally in reinforced thermoset molding materials. Combined with organic and inorganic fibers and fillers, the phenolics provide dimensionally stable compounds with excellent moldability.

Originally, phenolics were made in a one-stage process from the condensation of phenol and formaldehyde in the presence of a catalyst. The process involved a long curing time and the resins were difficult to control. A two-stage process is now used: the first stage consists of making a soluble and fusible resin called novalak, which is then converted into a thermosetting resin. The second stage phenolics are called resols, and by heating resols, a fully polymerized and infusible cresol is produced.

Phenolics are formed by compression and injection molding and extrusion. Injection molding usually provides the fastest cycle time, but it may not produce the best properties. For example, in long fiber filled compounds, compression molding gives the greatest strengths. Phenolic resins are used as bonding and impregnating materials. Bonding resins are available as pure compounds, but are more often formulated with elastomers and fillers to provide special properties. Phenolics are used in laminates, moldings, surface coatings, and adhesives. Laminates are usually reinforced with paper, fiber, or cloth.

Properties

Phenolic molding compounds are characterized by low cost, superior heat resistance, high heat distortion temperatures, good flame resistance, excellent dimensional stability, good water and chemical resistance, and excellent moldability. Phenolic compounds are classified as general purpose, nonbleeding, heat resistant, impact, electrical, and special purpose. Most compounds are black.

Applications

Wood flour-filled moldings are the lowest cost phenolic materials. Most are of the two-stage type and are used for products such as electric wall plates, industrial switch gear, circuit breakers, and handles for small appliances. Dimensional stability is improved by using mineral fillers, such as in power brake components. Nonbleeding compounds are one-stage odor-free materials used for cosmetic and drug containers.

100

Heat resistant compounds are made from two-stage resins and contain mineral fibers. Applications include motor housings for vacuum cleaners, handles for pots and pans, automotive transmission rings, and electrical components used at high temperature. Impact molding materials contain glass or other reinforcements. Typical applications are welding rod holders, thermostat housings, commutators and small motor housings, and heavy duty electrical components. Electrical phenolic components are used in small tool housings, automotive ignition housings, and components for aircraft and computers.

Special purpose molding compounds are formulated to meet specific properties, e.g., valve components which require dimensional stability at high temperature. Other applications of phenolics involve bonding, impregnation, and coating formulations. Phenolics are used in synthetic rubbers for tires and hoses in order to provide tacky surfaces. Solvent cements based on phenolics are used as structural adhesives for metal-to-rubber bonds and for bonding vinyls to various substrates. Phenolic resins are also used as heat resistant binders for brake linings, clutch faces, and other friction products.

Mineral wool and glass fiber insulation are other products that use phenolic binder resins. They are particularly suitable for humidity resistant applications such as in refrigerators. Phenolic compounds are also used in dip coatings for condensers and to impregnate paper stock for separators in automotive batteries. Oil, air, and fuel filters are also made from phenolic impregnated paper.

Polycarbonates

Polycarbonates possess very high impact strength, glass clear transparency, heat resistance, high dimensional stability, and chemical resistance. These properties make polycarbonates ideal for vandal-proof public lighting fittings, windows, and doors. Risk of breakage can be virtually eliminated by glazing with Lexan, a polycarbonate sheet several hundred times stronger than glass. It resists hammers, bricks, sparks, sprayed liquids, and heat.

Polycarbonates are also available in a variety of colors produced with pigments. They have good creep resistance and are excellent electrical insulators. Because of its high heat resistance (up to 145 °C) polycarbonates are one of the few thermoplastics suitable for use with high output lighting where temperatures can exceed 100 °C. They are also nonflammable and self-extinguishing. Under excessively high temperature, however, polycarbonate breaks down with the evolution of carbon dioxide gas.

Being resilient, the material resists denting. Its high rigidity, often without the need for reinforcing ribs, allows design freedom; foam molding allows it to be produced in many shapes. Structural foam polycarbonate has a strength-to-weight ratio between two and five times that of conventional metals. Typical applications include automotive interior panels and external bumper extensions, appliance and business machine housings, and bins for industry.

Solid opaque polycarbonate grades are widely used for high impact applications such as crash helmets, photographic equipment, and in electronics and electrical equipment. They are also used for the production of binocular bodies because of their breakage resistance and for water pumps because of their corrosion and abrasion resistance.

Polyesters

Although thermoset polyesters were first produced over 100 years ago, they have only been used widely for engineering components since World War II. Only recently have thermoplastic polyesters been introduced commercially for high performance injection molded parts.

Thermosets

Thermosetting polyester resins are low molecular weight alkyds crosslinked with a monomer (usually styrene) in a catalyst initiated reaction. At room temperature, the mixture of monomer and resin is stable for months. A variety of peroxide catalysts can initiate crosslinking at room temperature or above. In contrast to most other plastics, polyesters usually contain substantial amounts of other materials. Advantages and limitations apply to each type of compound and processing method.

Fabrication of parts from polyesters, which almost always are glass fiber filled, is more varied than with any other type of plastic. Methods of producing polyester/glass (FRP or GRP) parts vary from hand lay up and spray up for large parts to compression moldings for moderate sized intricate parts, and to cold press moldings for smaller components. Polyesters are also available as casting resins both as water extended formulations for low cost castings and as compounds filled with ground wood or nut shells. 'Low profile molding resins' are mixtures of polyesters, thermoplastic polymers, and glass fiber reinforcement and are used to produce parts with smooth surfaces that can be painted without need for prior sanding. Bulk molding compounds are mixtures of polyesters, glass fiber ranging from 3 to 6 mm in length, fillers, and a catalyst. Other additives can be used. These compounds are available in bulk form or, for ease of handling, as extruded rope. Sheet molding compounds consist of polyester resin, long glass fiber (up to 100 mm), a catalyst, and other additives. It is supplied in rolls, sandwiched between polyethylene films. Polyesters are also available for use as coatings which cure by ultraviolet radiation.

Outdoor weather resistance of polyesters is good for resins designed for such service. Thermal, electrical, and environmental properties are generally good. Major applications are for glass fiber reinforced structures such as boat hulls, architectural panels, and automobile and aircraft panels. Other uses include swimming pool filter bodies, athletic equipment, and stackable chairs.

Thermoplastics

Thermoplastic polyesters are high performance resins. Their molding compounds are crystalline, high molecular weight polymers that have excellent

102

properties and processing characteristics. Because they polymerize rapidly, mold cycles are short and molding temperatures are lower than for most thermoplastics. These polyesters are also available in glass reinforced compounds and in a talc-filled grade.

Thermoplastic polyesters have excellent resistance to chemicals at room temperature but are attacked by strong acids and bases. Dimensional stability is excellent and moisture absorption is low. Polyesters discolor when subjected to ultraviolet light for long periods. Black pigmented grades are recommended for outdoor applications for maximum strength retention. Toughness is another feature of thermoplastic polyesters. Unnotched specimens are rated 'no break' at room temperature and at -40 °C.

Although impact strength is excellent, polyesters are notch sensitive. They have outstanding electrical properties which are only slightly affected after prolonged exposure to high temperature or water immersion. For continuous exposure, water temperature should not exceed 50 °C.

Among the important uses for unmodified thermoplastic polyesters are housings which require good impact strength. The fact that polyesters produce hard smooth surfaces, are dimensionally stable, and exhibit low coefficients of friction allows them to be used in gears, cams, rollers, and bearings. Flame retardant types are used in televisions and other electrical and electronic components. Reinforced compounds are used in distributor caps and in painted exterior body components in cars. Darcon and Terylene are well-known trade names of polyester fibers.

Polyethylene and Chlorinated Polyethylene

Polyethylene

Polyethylene, a thermoplastic, was first discovered about 50 years ago. There are two types:

1. the low density or high pressure polyethylene which has a branched chain structure;
2. the high density or low pressure polyethylene which has a linear chain structure. Both types are obtained from natural or petroleum gas.

To produce low density polyethylene, ethylene is first obtained by refining the gas. It is then compressed, a peroxide is added as a catalyst and chain modifier, and hydrogen or carbon tetrachloride are also added. Polymerization occurs at elevated temperature, after which the polymer is extruded in ribbon form, cooled, and granulated. High density polyethylene is obtained from gaseous ethylene fed into an inert hydrocarbon solvent with a catalyst. Polymerization occurs at atmospheric pressure and at moderate temperature. The slurry output is then dried and the polymer separated.

Properties

Since high and low density polyethylene can be blended, a range of polyethylene grades is available with specific gravities from 0.91 to 0.96. Tensile strength, rigidity, and resistance to stress cracking increase with increasing specific gravity, whereas toughness decreases.

Low density polyethylene is a partially crystalline solid melting at 115 °C. It has a low specific gravity, flexibility without the use of plasticizers, good resilience, high tear strength, and good moisture and chemical resistance. High density polyethylene, being of higher crystallinity, is stiffer than low density polyethylene, with greater brittleness and higher strength (30 MPa as compared to 10 MPa). It has high resistance to environmental stress cracking. Polyethylene has excellent resistance to most acids and alkalis at normal temperature, although oxidizing acids will cause deterioration. At higher temperatures, polyethylene is soluble to varying degrees in hydrocarbons but is insoluble in liquids such as alcohols, vegetable oils, and most acids. It also has good impermeability to gases and very low water absorption rates.

Applications

Low density polyethylene is mainly used in the form of films and sheeting. Film thicknesses of 0.025 – 0.127 mm are extruded through small dies at 200 – 250 °C. Their greatest use is in packaging produce and perishable goods. Toys and household articles are injection molded, hollow sections such as pipes, ballpoint pen tubing, wire and cable insulation are extruded, and 'squeeze' bottles are blow molded. Polyethylene is also used as protective coatings, applied by dipping in hot solutions or emulsions or by spraying. Low density polyethylene is an ideal insulator because of its nonpolar nature and because it does not introduce electrical losses at high frequencies. Hence it is used in radar, television, and multicircuit telephone cables.

The main use of high density polyethylene is for housewares and toys produced by injection molding. Other uses include pipes, filaments for fabrics such as industrial cloths and filters, and ropes. The very high molecular weight types are used for industrial components which replace leather, wood, rubber, bronze, and steel.

Polyethylene tape is used for pipeline coatings to prevent corrosion. The bond created through the use of primer and polyethylene tape is resistant to the under-film migration of water, making it possible to lower freshly coated pipe directly into water filled ditches or swamps and other water crossings. Also, the glass transition temperature of polyethylene is less than −40 °C. Thus the tape can be applied at very low temperatures. The tape possesses soil resistant characteristics because the backing is smooth and nonadherent. Thus, the soil cannot pull the tape coating away from the pipe when the soil contracts or when there is relative movement between pipe and soil. The low moisture absorption and high electrical resistance of polyethylene are advantages in this application.

Another new application is that for porous sintered polyethylene given the name 'Nopol', which is used as a filtering material. Its major characteristics are an ability to absorb oil and petrol, silence compressed air, aerate liquids, filter particular matter, and repel water. Hence, it is used to soak up oil and gasoline in small boat bilges; as a grease trap; to reduce light oil pollution of rivers and waterways; as a silencer of exhaust noise on pneumatic hand tools, pneumatic control systems, and truck air braking systems; as a filter for air, gases, or liquids; and as water filters for compressor air intakes. It can be produced in rod, tube, sheet, or molded form with varying pore sizes.

Glass reinforced polyethylene is used for water-activated storage battery containers. The advantages are good impact resistance over a wide temperature range, outstanding heat resistance, excellent vibration resistance, superior strength properties, excellent resistance to both concentrated acid and dilute electrolyte, and low cost.

Chlorinated Polyethylene (CPE)

Chlorinated polyethylene is a thermoplastic resin which is crystalline in nature and has a high resistance to thermal degradation at molding and extrusion temperatures. CPE possesses a unique combination of mechanical and electrical properties and is extremely resistant to heat and chemical attack.

CPE, unlike most thermoplastics, maintains strength at reasonable elevated temperatures. When compared to other corrosion resistant thermoplastics, CPE has a higher creep resistance. CPE is resistant to more than 300 chemicals and reagents and is very useful in chemical plants for liners and coatings. It has a melting point of 180 °C, specific gravity of 1.4, and water absorption coefficient of 0.01%. CPE has good dielectric properties. Loss factors are somewhat higher than those of fluorocarbons, but lower than most thermoplastics. Its dielectric strength is also high and electrical properties in general vary little over a range of frequencies and temperatures.

CPE is used mainly as a corrosion resistant material at high temperature. Hence it is used in pipe fittings, tanks and processing vessels, valves, pumps, and meters. 'Penton' is one trade name of CPE.

Polyimides

Polyimides are a family of some of the most heat and fire resistant polymers known. Their excellent retention of mechanical and physical properties at high temperature is due to the nature of the aromatic raw materials used in their manufacture. Polyimides are formulated as thermosets and thermoplastics. Moldings, laminates, and resins are generally based on thermosets. Thin film products such as adhesives and coatings are usually derived from thermoplastic polyimide resins.

Laminates utilize continuous reinforcements such as woven glass and quartz fabrics, and boron and graphite fibers. Molding compounds contain chopped glass or asbestos or particulate fillers such as graphite powder. Polyimide films and wire enamels are generally unfilled. Coatings may be pigmented or PTFE-filled to give better lubrication. Adhesives generally contain aluminum powder to provide a closer match to the thermal expansion characteristics of metal substrates and to improve heat dissipation.

Polyimide parts are fabricated by techniques that range from extrusion and

105

powder metallurgy to injection, transfer, and compression molding methods. Generally, the greater the heat resistance of a polyimide, the more difficult it is to manufacture. Parts and laminates can operate continuously in air at 280 °C. They can withstand short exposures to temperatures as high as 500 °C. Glass fiber reinforced polyimides retain 70% of their flexural modulus at 250 °C. Creep is almost nonexistent even at high temperatures.

Polyimides have good wear resistance and low coefficients of friction. Both of these properties are improved by using PTFE fillers. Self-lubricating parts containing graphite powders have flexural strengths above 70 MPa, which is considerably higher than most thermoplastic bearing materials. Electrical properties are outstanding over a range of temperature and humidity conditions. Parts are unaffected by exposure to dilute acids, hydrocarbons, esters, ethers, and alcohols. They are attacked by dilute alkalis and concentrated organic acids.

Polyimide film has good mechanical properties to 600°C. An outstanding features is that, at 4 K, polyimide film can be bent around a 6 mm mandrel without breaking, and at 500 °C its tensile strength is 30 MPa. Polyimide adhesives maintain useful properties for over 12 000 h at 280 °C and for 100 h at 370 °C. Resistance of adhesives to combined heat and salt water is excellent. Molded glass reinforced polyimides are used in jet engines and in high temperature electrical connectors. High speed, high load bearings for business machines and computer printout terminals use self-lubricating polyimides. Other applications include bearing cages for gyroscopes and air compressor piston rings.

Polyphenylene Oxides

Polyphenylene oxides (PPO) are thermoplastic materials. Their structure is such that by allowing chains to rotate, but not bend, good mechanical and thermal properties are achieved.

Properties

Among engineering plastics these materials are remarkable for their dimensional stability. They have very low water absorption, low coefficients of thermal expansion, high heat distortion temperatures, and excellent resistance both to creep under continuous load and to fatigue. PPO is nontoxic and self-extinguishing and, with a specific gravity of 1.06, is a very lightweight thermoplastic. Modified polyphenylene oxides such as glass-filled grades are used in parts subjected to prolonged stresses. Impact resistance is excellent. Polyphenylene oxide is processed by extrusion and injection molding. It is available in transparent and opaque formulations.

Applications

Both filled and unfilled grades are used frequently where high performance is required without excessive cost. Some reinforced grades compete with diecast metals. Typical applications include portable household or kitchen appliance housings such as hair dryers and dishwashers, radio and television components, waveguides, office machine housings, water distribution system parts, electronic connectors, automotive accessories, and surgical instruments. High strength-to-weight ratio foams have been made with PPO.

References

CRAIG, A.S. *Dictionary of Rubber Technology*, Butterworths, London, 1969

ALLEN, P.W. *Natural Rubber and the Synthetics*, Crosby Lockwood, London, 1972

ROFF, W.T. and SCOTT, J.R. *Fibers, Films, Plastics and Rubbers*, Butterworths, London, 1971

Polypropylene

Polypropylene is a relatively new thermoplastic, having been commercially available for only 20 years. The propylene monomer is produced by the 'cracking' of petroleum products such as natural gas. With the aid of catalysts such as titanium trichloride and diethyl aluminum chloride, polymerization is carried out to form a slurry with $80 - 85\%$ polypropylene. The mixture is then washed, dried, and granulated. Although polypropylene exists in three forms, the isotactic form in which the ordered methyl groups stiffen the linear chain is the most useful as it has increased melting point and crystallinity. The qualities of polypropylene may be improved by adding materials before polymerization; e.g., ethylene to toughen the polymer, carbon black to reinforce it, glass fiber to improve stiffness, and rubbers such as butyl to reduce brittleness.

Properties

Polypropylene is quite similar in properties to polyethylene. It is, however, the lightest thermoplastic known and has a specific gravity of 0.9. It has a high melting point of 150 °C and a tensile strength of 35 MPa. The tensile strength, rigidity, and crack resistance are greater than polyethylene. Polypropylene has good abrasion resistance, excellent electrical properties, and high creep resistance. It is resistant to most acids, alkalis, and saline solutions even at high temperatures. However, polypropylene is soluble in substances such as toluene and chlorinated hydrocarbons. Being the lightest thermoplastic and having high crystallinity, it has a high strength-to-weight ratio. At temperatures of about 0 °C it is brittle compared with high density polyethylene and is more vulnerable to oxidation.

Applications

Injection molding of polypropylene produces a wide variety of products.

Since it possesses rigidity, toughness, and heat resistance, it is used in hospitals as it can be sterilized to 135 °C for long periods without severe damage. It is ideal for hinged containers; for example, polypropylene soap box hinges can be flexed several thousand times without suffering fatigue damage. Injection molding produces automotive parts such as shelves, seals, and steering wheels as well as chairs, television cabinets, washing machine tubs, tops, and lids, dishwasher and vacuum cleaner parts, furniture and luggage. Air ducts and tanks can be blow molded. Extruded products are wire insulation, sheets for chemical plants, and ropes. Polypropylene is used for encapsulating capacitors due to its electrical insulating properties and because it does not suffer from environmental stress cracking.

Asbestos reinforced polypropylene is finding application in the automobile industry as heater and air filter casings. Propathene, a glass reinforced polypropylene compound, is suitable for housings for water pumps and for lamps. Its tensile strength at 135 °C is similar to that of unfilled polypropylene at 20 °C. Glass filled polypropylene is also used for cooling fans, small carburetor parts, and water pump impellers.

Polystyrene

Polystyrene was first produced in 1831 but has only become commercially significant over the last 50 years. Present consumption is measured in millions of tonnes per year. Polystyrene is made industrially in large quantities because equipment is expensive and the process requires several days between intake and discharge. Rigid benzene under pressure is added to ethylene in the presence of aluminum trichloride to produce ethyl benzene. The product is reduced to the styrene monomer by passing it over an iron oxide catalyst at high temperature. Free styrene is then mixed with a peroxide and the resulting polymer is passed through a cylindrical tower where the reaction is controlled by heaters. It is then extruded and granulated.

Polystyrene is available in many forms and grades but there are three main groups:

1. general purpose or unmodified polystyrene, i.e., polymerized styrene with or without additives but excluding rubber and copolymers
2. toughened or high impact polystyrenes which include impact strength ranges (these consist of styrene polymerized or mechanically blended with rubber)
3. styrene – acrylonitrile copolymer

Properties

Like most polymers, polystyrene has good chemical resistance. It is resistant to alkalis and acids, hydrochloric and sulfuric acids, and oxidizing and reducing agents. Its outstanding property is ease of processing. Polystyrene flows easily, has minimal shrinkage, low moisture absorption, and good dimensional stability and is readily injection molded. It has good electrical insulating properties, low thermal conductivity, is easily colored, and has a moderate tensile strength. Although optically transparent, prolonged exposure to sunlight causes yellowing of polystyrene. Its brittleness and elevated temperature performance limits its usage. Polystyrene is attacked by many solvents – for example alcohols, oils, and dry cleaning agents – and crazes on exposure to weathering. General purpose grades have good heat resistance. High molecular weight grades have improved strength. Heat resistant grades have softening points from 70 to 100 °C.

Applications

One-third of the total packaging market consists of vending cups thermoformed from extruded and toughened polystyrene sheet. Apart from cost, rigidity, lack of taint, and satisfactory temperature resistance, toughened polystyrene is favored because it can be formed. Special easy-flow grades of toughened polystyrene have been developed for injection molding of thin-walled containers for dairy products and foodstuffs. Crystal polystyrene is used similarly for cheeses and jams, the latter requiring temperature resistant grades to allow hot filling.

Lavatory cisterns use large quantities of toughened polystyrene, mainly in white and light colors. Wall tiles and bathroom cabinets are made by injection molding and by thermoforming; large bath panels are made from thick polished sheet by thermoforming. The use of polystyrene in furniture drawers, legs, and other components is increasing slowly but could become a future major usage. Nearly all domestic refrigerators have liners made from thermoformed sheet. Commercial refrigerators commonly have capping strips and door components of the same material. Toughened polystyrene is also widely used in vacuum cleaners, cameras, projectors, televisions and radios, record players, tape recorders, and in spools, cassettes, and catridges for tape and photographic film.

It has been suggested that expanded polystyrene foam could be used for insulating roads in countries such as Scandinavia which suffer from severe winters. In these countries, roads experience 'frost heave' due to frost penetration as deep as 1.5 m. If insulating materials were used in road construction, it would be expected that the rate of heat loss from under the road would be reduced considerably.

A new thermal insulation material is self-extinguishing expanded polystyrene sheet held within a black polyethylene film envelope. The envelope protects the core and provides an effective lapping to produce watertight joints. The material may compete with roofing felt and glass fiber insulations.

Another major building use of expanded polystyrene is for forming voids in concrete to reduce the amount used. Previously metal tubes and wooden shuttering were used, but by employing polystyrene significant cost reductions have been possible.

Polystyrene foam has been used for lagging pipelines carrying chemicals at temperatures of − 100 °C. It is ideal for high performance lagging because, in addition to a low coefficient of thermal conductivity, it is easy to fabricate and is

dimensionally stable. Unusually low absorption and permeability characteristics of the foam suit applications involving flammable liquids.

References

BRYDSON, J. *Plastics Materials*, Iliffe Books, London 1966
BILLMEYER, F. *Textbook of Polymer Science*, Wiley Interscience, New York,
1965
Principles of High-Polymer Theory and Practice, McGraw-Hill, New York,
1948
SEYMOUR, R. *Introduction to Polymer Chemistry*, McGraw-Hill, New York,
1971
Engineering Materials and Design, December 1970, June 1972, September 1972,
May 1974, IPC Industrial Press

Polysulfone

Polysulfone is one of the newer types of plastic, having been introduced in
1965. It is a strong, stiff engineering plastic noted for having the highest heat
distortion temperature of the melt processable thermoplastics. These resins
must be dried for several hours at 120 °C before being molded. The natural
color of polysulfone is transparent light amber; resins are also available in transparent and opaque colors.

Properties

The heat distortion temperature of polysulfone is 170 °C. Stiffness, thermal
stability, and oxidation resistance of molded polysulfone parts remains high at
service temperatures to 160 °C. Even after long exposure to such temperatures,
the resin does not discolor or degrade. Polysulfone is tough, and unnotched
impact specimens do not break. However, the material is notch sensitive and
notched impact values are only moderate.

Strength, stiffness, and electrical properties of polysulfone are excellent in
long-term performance. Dimensional stability is excellent under severe environments and creep is low even at elevated temperature. Polysulfone has excellent
resistance to solutions of inorganic acids and alkalis. It is attacked by solvents
such as esters and ketones and is soluble in chlorinated hydrocarbons.

Applications

Electrical applications of molded polysulfone include coils, bobbins, connectors, integrated circuit carriers, and housings. Glass filled polysulfone is used
in 'under the hood' automotive components such as switch and relay bases.
Appliance parts include coffee makers, humidifiers, high intensity lamps, steam

110

irons, and hot water tanks for beverage makers. Extruded sheet and film are used for overhead projector transparencies and in heat fusing copying machines. Polysulfone parts are not recommended for outdoor use unless they are painted or plated. They may be sterilized by wet or dry heat or by gases and are used in medical instruments, hospital breathing apparatus, and ultra-filtration equipment.

References

Engineering Materials and Design, Volume 18, Number 4, May 1974 and
   Volume 18, Number 8, September 1974, IPC Industrial Design
DUBOIS J.H. and JOHN F.W., *Plastics*, Reinhold, New York, 1967

Polyvinyls

Polyvinyls include polyvinyl acetate, butyral, chloride, chlorinated PVC, and vinyl copolymers.

Polyvinyl Acetate (PVA)

Polyvinyl acetate was first prepared in 1912 and was subsequently put into large-scale production in Germany and in Canada. Initially PVA was made by passing acetylene into acetic acid in the presence of a catalyst. More recently, it has been derived less expensively from the reaction of ethylene, oxygen, and acetic acid with a metallic catalyst. Vinyl acetate has been polymerized in bulk, in solution, in suspension, and in emulsion. Of the material identified as PVA or copolymers, 90% is made by emulsion techniques.

PVA base products are used primarily in adhesive and paint applications, the main uses being packaging and wood gluing. The emulsion form of PVA is especially suitable for adhesives because they are stable and accept many types of additives without being damaged. Adhesives can be adapted to many types of machine application, from spray guns, rollers, extruder type devices, and high speed gluing equipment. They are excellent bases for water resistant paper adhesives in manufacturing bags, tubes, and cartons. Aqueous dispersions are also used as adhesives and lacquers for leather, water, and grease proofing, sealing, and paper packaging.

PVA emulsion paints form flexible, durable films with good adhesion to clean surfaces including wood, plaster, concrete, stone, brick, blocks, asbestos board, asphalt, tar paper, wall boards, aluminum, and galvanized iron. Their ability to satisfy the two requirements of any paint—to provide protection and decoration—is supplemented by several further advantages: they do not contain solvents, so that physiological harm and fire hazards are eliminated; they are odorless, easy to apply, and dry rapidly. They may be thinned with water, and brushes or rollers can be cleaned with soap and tepid water.

The durability of PVA paints, particularly their resistance to chalking, surpasses that of conventional oleoresinous paints. PVA films have high resistance to degradation by ultraviolet light. Paints that are correctly formulated to form quality latexes develop little or no chalk, thus giving maximum tint retention and endurance. Blister resistance of PVA paints is another important advantage for exterior use. Latex paint films are permeable enough to permit water vapor, but not liquid, to penetrate them, which prevents blistering and peeling. Film formation is not impaired if they are painted on damp surfaces or applied under very humid conditions.

Other uses of PVA emulsions and resins are as binders in coatings for paper and paperboard. Emulsions find wide application as textile finishes because of their low cost and good adhesion to natural and synthetic fibers, leather, asbestos, sawdust, sand, clay, etc., to form compositions which may be shaped with heat and pressure. The compressive and tensile strength of concrete is improved by addition of PVA emulsions to the water before mixing. Such formulations also aid adhesion between new and old concrete when patching and resurfacing. Vinyl acetate homopolymer and copolymer have been used for chewing gum bases.

Solid PVA has a melting point of 60 – 88 °C and an average specific gravity of 1.18. Because of its chemical composition, PVA can be recognized by flame testing—it burns with a yellow smoky flame, giving off an acetic odor. The gas is acidic and can be detected by moist litmus paper. The plastic loses dimensional stability at temperatures exceeding 38 °C. Above 55 °C it becomes rubberlike and at greater temperatures it becomes tacky.

Polyvinyl Butyral (PVB)

Polyvinyl butyral, another thermoplastic, is formed from butyraldehyde which reacts with polyvinyl alcohol. It is used as an interlayer for safety glass as well as for coating fabrics and for injection molding compositions. The resin is usually prepared containing large numbers of hydroxyl groups which act as plasticizers and make the resin adherent to glass. The properties of PVB are similar to those of PVA:

> clarity in thin films
> dielectric constants are similar
> good water resistance
> heat distortion temperatures are low
> inexpensive
> outstanding adhesion to a wide variety of materials
> specific gravities are identical
> stability to ultraviolet light
> water absorption and specific heat are similar

Films can be cast from solution or extruded and are sealable at 200 °C. For surface coatings, the polymer is applied from solution often with other resins (e.g., shellac or cellulose nitrate). For maximum mechanical and thermal endurance, surface coatings and structural adhesives are based on PVB and a crosslinking component – butyral and phenol – formaldehyde provide a desirable stoving enamel. Plasticized PVB can be applied (e.g., to textiles) from aqueous dispersion. PVB can be compression molded at 120–165 °C or injection molded at 150–185 °C.

Utilization is mainly as can coatings, waterproofing natural and synthetic textiles, and as vacuum bag films. PVB is used as the interlayer for automotive safety glass because of its toughness, flexibility, and outstanding adhesion. In dispersion form, its main applications are in textile treatment, strip coatings, hot melt adhesives, and wash primers.

Polyvinyl Chloride (PVC) and Chlorinated PVC

PVC, second only to polyethylene, is the most common of all vinyl polymers. The polymer of vinyl chloride, PVC, is produced by three basic processes:

1. *Mass process*. Vinyl chloride is polymerized in a carrying medium. The polymerization is stopped at a low conversion and the polymer is separated from the residual monomer which is recycled. The product is a granular material.
2. *Emulsion process.* Vinyl chloride is emulsified with water by use of emulsifying agents. The polymerization is carried out to a high degree of conversion with a small amount of recovered monomer being recycled. The product is obtained as an emulsion or spray, and dried to produce a very fine powder.
3. *Suspension process*. Vinyl chloride is suspended as small droplets of monomer in water and the polymerization is carried to a high degree of conversion. A small amount of monomer is recovered and recycled. A granular product is obtained by centrifuging and drying.

While a range of products can be made by the first two methods as above, they are generally limited to speciality products which are difficult or impossible to make by the suspension process. The largest proportion of PVC is produced, particularly in the USA, by the suspension process because it requires the least capital investment and has the lowest operation cost.

Properties   Combustion of PVC softens and chars the material. It burns, but is self-extinguishing unless combustible additives are present. When kept burning in a bunsen flame, it produces a very sooty yellow flame with a green tip and leaves a black residue. It gives off a bright green flame with the copper wire test.

PVC combined with plasticizers (e.g., high boiling organic esters), lubricants, fillers, and stabilizers form a wide versatile range of compounds from soft flexible films to rigid high strength products. PVC can be divided into two categories: rigid and plasticized. From these it is possible to make a compound for almost any application. The rigid form usually has little or no plasticizer added. Plasticized PVC is softer and flexibility is dependent on the type and amount of plasticizer added. Compounding ingredients include stabilizers, lubricants, fillers, pigments, processing acids, and modifiers.

Polyvinyl chloride has attained market leadership because of its good physical properties, its compounding versatility, its low cost, and its ease of processing. It has good water, chemical, and abrasion resistance, high strength-

113

to-weight ratio, and is available in a complete range of colors. Good weather-ability is achieved by proper compounding PVC is resistant to most acids, fats, petroleum, salts, and fungus growth. Its disadvantages are that rigidity varies markedly with temperature, it decomposes at 100 °C, and it is attacked by ketones and esters.

Uses Principal uses for PVC include pipe and fittings, wire and cable insulation, weather stripping, rain guttering, and septic drainage. PVC pipes can be easily cut with a hacksaw and joined with gluing compounds, are very light, and can be subjected to a reasonable degree of bending. Ordinary low cost PVC cannot, however, be used for hot water plumbing.

Flexible or plasticized PVC is used for raincoats, toys, bottles, shoe soles, wire insulation, film sheeting, garden hose, gaskets, closures, shower curtains, and inflatable toys. PVC is a desirable blow molding material because of its clarity and fracture resistance. Toiletries, cosmetics, household chemicals, and food products are contained in PVC bottles.

A recent development in PVC applications is High Tenacity Trevira. It is a PVC coated polyester fabric which can be welded or stitched. The fabric is used for flexible containers and tarpaulins.

Chlorinated PVC resin is a hard and rigid thermoplastic but it can be softened to appear leather-like by processing with chemical agents. It is self-extinguishing and will not burn when exposed to high temperatures. CPVC has a higher tensile strength than most thermoplastics and is rot proof. It has excellent corrosion resistance to alkalis and weak acids and high stability to water and mineral acids, but moderate to poor resistance to stronger acids. It will absorb some water. It has no resistance to ketones and esters. CPVC is used mainly in corrosive environments at high temperatures because of its excellent chemical properties and heat resistance and in pump part lining, in valves, heat exchanger pipes, and chemical processing equipment. It is also used in hoods, ducts, and exterior building components. Another recent development is that of polyvinyl fluoride (PVF), an extremely tough and highly extensible film used in exterior protection of homes and buildings.

Vinyl Copolymers

By varying the ratio of vinyl chloride to vinyl acetate, resins with specifically desired properties can be prepared. Polymerization of the mixture gives a chainlike molecular skeleton of polyvinyl chloride having chloride and acetate groups attached. The major types of copolymers are:

1. those with a vinyl chloride-to-vinyl acetate mass ratio of about 85:15, which gives a rigid material for sheet stock and molding
2. those with vinyl chloride up to 95%, which yield flexible forms for sheeting
3. vinylidene chloride
4. acrylonitrile and vinyl stearate
5. polyvinyl acetate butyrate

114

Copolymers are generally compounded with a thermal stabilizer and lubricants, coloring agents, and inert fillers to produce molding powders which are transparent, translucent, or opaque. Objects molded from them are the most dimensionally stable of all plastics. Nonflammability, durability, and water resistance far exceed those of pure vinyl polymer. However, the copolymers suffer a decrease in softening temperature, hardness, and stiffness.

Vinyl chloride acetate copolymers are used for high fidelity records, because of excellent groove reproduction from the mold. The copolymer is extensively used for floor products because of the large amount of filler it will bond with, while retaining good colorability and physical characteristics. Copolymer vinyl is calendered into flexible or rigid sheets for products requiring better dimensional control, heat stability, and abrasion resistance.

Vinylidene chloride is an important packaging and industrial material. It is tasteless, odorless, nonflammable, tough, and abrasion resistant. It displays high tensile strength; monofilaments are used as screens and high tensile fabrics. It has impermeability to many flavorings. Vinylidene chloride plastics are among the most inert of all thermoplastics. Principal applications include auto seat covers, upholstery, draperies, carpeting, food packaging, film, bristles, pipe and pipe linings, fillers, awnings, paper and paper board castings, valves, and pipe fittings.

Copolymers with acrylonitrile make excellent coating materials for other plastics where they add impermeability and grease and oil resistance. All vinyl materials are used below the boiling temperature of water and are resistant to reagents but are soluble in ketones.

References

KIRK, R.E. and OTHMER, D.F. *Encyclopedia of Chemical Technology*, 3rd edn, John Wiley, New York, 1978
SHELL, F.D. and HILTON, C.L. *Encyclopedia of Industrial Chemical Analysis*, Interscience, New York, 1966

Silicones

The silicones are a family of polymeric organic compounds produced in a vast variety of forms, ranging from light lubricating oils to structural plastics and space vehicle heat shields. Although research into silicones began in the early 1900s, products did not become commercially available until the Second World War. Today, silicones are used in a large number of industrial and domestic applications including fluids, greases, compounds, resins, rubbers, and elastomers.

Chemically, silicones are polymers comprising alternate silicon and oxygen atoms linked to organic molecules. Production is almost universally by direct synthesis, which involves various silicone compounds being drawn off at different stages, similarly to the distillation of petroleum products from oil. The raw materials used in silicone production are silicon (quartz sand) and carbon in the form of coke or charcoal.

Silicone Fluids

Silicone fluids may be divided into several classes according to chemical structure, but all possess the same general properties which enable specialized usage. The most important property which these fluids possess is their very wide service temperature range: freezing points vary from -22 to -86 °C, while the upper continuous use service temperature is 250 °C. At elevated temperature, the fluids are chemically stable and suffer no oxidation. They do not readily burn, and they self-extinguish once the combustion source is removed. A property unique to the silicone fluids is that viscosity is not temperature dependent, the range of available viscosities being very wide. Fluid compressibility is remarkable with volume changes of up to one-third possible. Water repellent properties are excellent with the fluids providing the protection afforded by petroleum oils, but with greater permanency.

Chemical inertness is excellent and the fluids have low toxicity. Low surface tension leads to the fluids being difficult to confine. The fluids are used as defoaming agents for many liquids. Silicone fluids possess superior lubricating properties and dielectric properties are excellent. Major applications of silicone fluids exploit their elevated service temperatures together with some other property: in power transmission fluids, it is dielectric properties; in vibration dampers and liquids springs, it is viscosity and compressibility; in lubricants, it is lubrication ability.

Their water repellent properties are used in coatings on all materials where water penetration is undesirable: on industrial equipment, fibers, and in furniture polishes. They are used as release agents in metals and plastics, molds, and cooking receptacles, while their use as defoaming agents is utilized in industrial, medical, and veterinary applications.

Silicone Greases

Silicone greases are fluids thickened with a filler, usually silica. Their properties are similar to those of the fluids. Service temperatures are from -75 to 250 °C combined with chemical inertness. Water repellency is 100% and the greases and nontoxic. They do not oxidize or solidify. Their lubrication properties are excellent for dissimilar metal combinations such as metal/rubber/plastic combinations. Grades are available to cope with light or heavy loading conditions. Other than high costs, the only disadvantage silicone greases have is that they can be used with steel/steel surfaces for light loads only. The greases otherwise match hydrocarbon greases in all conventional applications and surpass them in all corrosive and high temperature environments. In lubrication conditions involving very high temperatures and loadings, they are surpassed only by molybdenum disulfide or PTFE solid lubricants.

Silicone grease lubrication is used for gear assemblies, ball, sleeve, and thrust bearings, conveyors, monorails, and industrial machine parts. They are also used in anti-seize thread compounds for easy assembly and disassembly, for pump and valve lubrication, and in chemical plants. In domestic applications, silicone greases are incorporated into conventional lubricants for use as light duty greases such as are used on fishing reel gear trains.

116

Silicone Compounds

Silicone compounds are substances having little fluidity. They are not solids, but are putty-like and do not melt. Major properties are high dielectric strength at temperature ranges comparable to the fluids, durability in outdoor environments, water repellency, chemical inertness, and high thermal conductivity.

Silicone compounds are used in the electrical industry as coatings on electrical connections, switchgear, and insulators. They greatly improve performance by protection from tracking and arcing. Other uses include coatings to prevent metal corrosion, as flexible retainers for parts such as rubber, plastic, and leather diaphrams, and as an insulating and water proofing filler in cable splices. Their high thermal conductivity ensures efficient heat transfer from power electronic components to heat sinks.

Silicone Resins

Silicone resins have the surface appearance of organic varnishes. The main types may be classified into coatings, moldings, laminates foams, encapsulants, and adhesives.

Coatings offer complete water repellency, have service temperatures to 550 °C, and are chemical inert. They are also nontoxic, quick drying, available in any color, and have excellent adhesion. They are applied to surfaces which require protection from heat deterioration, water penetration, or chemical attack.

Silicone resin moldings are thermosets which can be formed by injection, compression, or transfer molding. They possess excellent electrical insulating properties, chemical inertness, and have elevated service temperatures greater than 270 °C. They may be reinforced by many fillers such as chopped glass fibers, granular silica, asbestos, and nylon fibers; however, tensile strength is not high, averaging about 42 MPa. Typical applications are for insulators, switches, high frequency electrical equipment covers, arc suppressors, and electronic components.

Silicone laminates exhibit outstanding mechanical properties. Tensile strength is very high, up to 240 MPa for woven glass fabric reinforced laminate. Maximum continuous service temperatures are to 250 °C. The laminates are chemically inert and possess electrical properties similar to those of silicones alone. Under extreme temperatures, silicone laminates break down layer by layer, which, if exposure time is short, allows them successfully to protect valuable components. This process of material deterioration is called 'ablation' and is used to protect space re-entry vehicles and rocket engine thrust chambers from temperatures to 8000 °C. Apart from space applications, the laminates are used in structural components subjected to elevated temperatures.

Silicone resins may be converted into low density foams (100 kg/m^3), have a maximum service temperature of 250 °C, and provide excellent thermal insulation. An added advantage is that foam can be formed in place. Uses are principally in the aerospace industry. Silicone resin adhesives are pressure sensitive, have high temperature resistance, and have excellent electrical properties.

117

Rubbers and Elastomers

The outstanding property of silicone rubbers and elastomers is their service temperature range of from −90 to 250 °C, with certain grades being usable to 300 °C. Electrical properties are excellent and service lives are long. They are chemically inert and fire resistant. Silicones decompose to products which still retain initial electrical properties which is important, for example, in electrical cable sheathing and encapsulation. Tensile strength of the rubbers is typically about 9 MPa, which is only fair when compared to other rubbers. Some varieties of silicone rubber, as with the laminates, undergo ablation under extreme temperatures.

Rubbers and elastomers may be molded, extruded, or fabricated using conventional techniques or they may be cast and cured in place. Typical applications include gasket and seals, shock mounts, vibration dampers, caulking and sealing compounds, and molds and adhesives. Striking examples of the harsh service conditions that silicone rubbers can withstand are elastomeric masking tapes, which resist temperatures of 450 °C, hydrochloric acid, and the extremely high temperatures of plasma arc spraying.

Unfortunately, although silicones possess many outstanding properties, they are all expensive and are usually used only where service conditions are too severe for conventional materials. There are, however, many industrial and domestic products which contain proportions of silicones to improve properties of the main constituent material and which are considerably cheaper. Examples are many aerosol spray products, caulking compounds, and light greases.

Ureas

Ureas and melamines make up the bulk of the family of plastics known as amino resins. They are thermosetting resins in which the filler material is almost always cellulose. Ureas data back to 1884; however, it was 1918 before production became significant. Raw urea used to manufacture resins is produced by a reaction between ammonia and carbon dioxide followed by dehydration. It is a white crystalline solid which, apart from being used to produce urea − formaldehyde resins, is an important agricultural fertilizer, being broken down in soil to ammonium carbonate.

Production of urea resins is based on the reaction between raw urea and formaldehyde, from which methyl compounds are produced by condensation. At this stage the product is water soluble and filler material is introduced by a vacuum process. Water loss initiates polymerization and subsequent crosslinking. After charging, the product is ground by ball mill to various grades according to particle size. The filler for resins, α-cellulose, is obtained from sulfate-bleached paper. Urea resins may be classified into four types: molding resins, adhesives, coatings, and foams.

Molding Resins

Urea − formaldehyde molding resins with cellulose fillers are produced by

118

compression molding since flow characteristics do not allow transfer or injection molding. Mechanical properties of the molding resins could allow ureas to be described as 'middle of the road' plastics. Tensile strength is as high as 70 MPa, higher than that of phenolic resins, which are major competitors. Impact strength is fairly low, about one-fifth that of the alkyds. Heat resistance is low compared to most resins, the maximum continuous service temperature being 80 °C, and the heat distortion point being 140 °C.

There are, however, notable properties of ureas. Surface hardness is high, as is mar resistance. The resins are absolutely transparent but can be dyed to virtually any color. Color fastness is extremely good. Complete chemical resistance to organic solvents is a valuable property. Electrical properties are comparable with those of good insulating materials, provided temperatures are below maximum service temperatures. Urea resins possess better than average properties compared with other plastics at room temperature. However, they do not match performance of the melamines.

The main applications for urea molding resins are domestic, especially in the kitchen. Tableware, stove and refrigerator hardware, and many appliance housings are made from urea or melamine resins. In these applications, where products are subject to considerable handling and wear, the most important property is surface hardness. In other parts of the home they are used for light fittings, electrical junction boxes, container closures, toys, and buttons. Urea molding resins are utilized in industry for housings and switchboard components in power supply circuits, in binding sand cores for metal castings, and as stabilizers in explosives.

Adhesives

Urea – formaldehyde resins are widely used in the plywood and furniture making industries as adhesives. Urea and melamine adhesives production accounts for one-third of total amino resin consumption. Fillers are added to the resin to adjust shrinkage and absorption properties suitable for the materials being joined. The adhesives are cured by application of an acid, which converts the low molecular weight resin into a high molecular weight polymer. They are two-part adhesives, with the resin a dry powder and the acid a fluid hardener. There are two methods of application. First, the acid may be mixed directly with the resin and applied to the two parts being joined, but here pot life is short and any surplus is usually wasted. Alternatively, the resin may be mixed with water and the resin mixture applied to one part and the acid catalyst to the other. Pot life in this case is longer, up to several days.

Properties of the urea adhesives include waterproofness, boil resistance, and very high surface tension. The latter property results in extraordinary gap-filling and bridging capabilities not found in other adhesives. They possess excellent resistance to low temperatures, fungal attack, and solvents. Melamine adhesives have better mechanical properties than ureas, but are usually used in urea – melamine combinations for reasons of economy. Urea adhesives are used mainly for furniture, cabinets, and plywood, but find some usage also in the paper and leather industries.

119

Coatings

Urea – formaldehyde resins are used as coatings which require high surface hardness, corrosion resistance, or very good color fastness. They are utilized in two forms. They may be applied directly to a surface and then cured, usually at elevated temperature and with the addition of a catalyst. Alternatively, solvent-thinned resins may be painted on to the surface. Often urea resins are combined with alkyds and used as baking enamels, which resemble vitreous enamels.

Urea coatings have both desirable and undersirable properties. They have high surface hardness and color retention. They are chemically resistant to acids, alkalis, solvents, and fungi. On the other hand, flexibility of the coating is low and for many coatings elevated curing temperatures are required. Applications of coatings include kitchen equipment, automobile parts, collapsible tubing, hospital equipment, refrigerators, and venetian blinds. They are also used as furniture lacquers and other timber finishes and as coatings for textiles such as cotton and rayon to impart crease resistance, stiffness, shrinkage control, fire retardance, and water repellency. When used to coat paper they improve wet strength, rub resistance, and bursting strength.

Foams

Urea – formaldehyde resins may be processed into extremely lightweight foams which may be used as an insulating layer (both thermal and acoustic) for instrument packing. Its thermal conductivity is the lowest of conventional resin foams. It is the lightest rigid foam available (5 kg/m^3) with a density about 50% that of polystyrene foam. The maximum service temperature of urea foams is only 55 °C. It has practically no tensile strength and it has a very low compressive strength of about 30 MPa.

References

Engineering Materials and Design, October 1975, p. 46; January 1976, p. 40; February 1976, p. 13
Electrical Review, 12 March 1976, Volume 198, Number 10, p. 32

8
Rubbers and Elastomers

The natural rubber molecule is said to be stereoregular, i.e., the molecule geometry is defined precisely in three dimensions. Until the 1950s efforts to achieve the stereoregular molecule in an artificial polymerization process had failed, but these efforts did result in the development of several nonstereo-regular synthetic rubbers. Rubbers falling into this group are termed 'first generation synthetics'. Some comparative properties of these and rubbers developed more recently are given in Table 10.

All synthetic rubbers utilize the same general manufacturing process. Their production requires the synthesis of monomers, or small molecules, almost invariably from petrochemicals, followed by polymerization to form long chain molecules. For example, butadiene – styrene requires two monomers for production, styrene—a liquid, and butadiene—a gas at normal temperatures; upon polymerization, the monomers combine to form the polymer, butadiene – styrene. Catalysts are required to initiate polymerization and, after initiation, polymerization takes place continuously.

Emulsion polymerization is the most favored process used, with the monomer dispersed in water as an emulsion. The process leads to the aqueous suspension of raw rubber particles called latex. The latex is coagulated to form solid granules of raw rubber. It is usually performed in two stages. First, a brine solution is added and the mix is then flocculated with sulfuric acid or aluminum sulfate. Second, the fine raw rubber granules are washed and the rubber is dried by vacuum filtration. Manufacture follows in the following sequence: (1) initial preparation, (2) mixing and blending with vulcanizing and other chemicals, (3) shaping and fabrication, and (4) vulcanization.

Initial preparation involves cutting up bales for the mixer and sometimes heating them. Carbon black, sulfur, and other vulcanizing and protective chemicals are added during mixing. The mix is then shaped by extrusion or molding. Vulcanizing produces crosslinking of the polymer chains. Without crosslinking, rubbers are thermoplastic and lack elasticity.

Butadiene Acrylonitrile

Butadiene acrylonitriles date back to the 1930s. They are commonly referred to as 'nitrile' rubbers, or NBR, and are classed as specialty rubbers. Their cost excludes them from large volume production. Nitrile rubbers are manufactured

Table 10 Polymer Selection and General Service Guide

	Natural rubber (natural polyisoprene)	Synthetic natural (synthetic polyisoprene)	SBR (styrene butadiene)	Butyl (isobutylene – isoprene)	Polybutadiene (polybutadiene)	EPR (ethylene – polypropylene)	Neoprene (polychloroprene)
	Common name and chemical composition of base elastomer						
Physical properties							
Durometer hardness range (Shore A)	30 – 95	30 – 95	30 – 95	30 – 90	40 – 95	30 – 90	20 – 95
Tensile strength (MPa)	30	30	24	20	20	20	28
Maximum elongation (%)	650	650	600	650	650	600	600
Compression set	E	E	G	G – F	G	E – G	G
Resilience	E	E	G	F	E	G	E
Gas impermeability	G	G	F	E	G	G	G
Electrical resistivity (polymer)	E	E	E	E	E	E	F
High temperature service (max.°C)	105	105	110	125	105	150	125
Low temperature service (Min.°C)	– 50	– 50	– 50	– 45	– 60	– 55	– 40
Mechanical properties							
Impact strength	E	E	E	G	G	G	G
Abrasion resistance	E	G	E	F	E	G	E
Tear resistance	E	E	F	G	G	F	G
Cut growth	E	E	G	E	F	G	G
Bonding to rigid materials	E	E	E	G	E	F	E – G
Resistance ratings							
Weather—sunlight ageing	F – P	N	F – P	E	F – P	E	G
Oxidation	G	G	F	E	G	E	E
Ozone cracking	N	N	N	E	N	E	E
Radiation	G	F	G	G	P	G	G
Water	E	E	E – G	E	E	E	G
Alkali—dilute/concentrated	E/G	G/G	G/G	E/E	G/G	E/E	E/E
Acid—dilute/concentrated	E/G	G/G	G/G	E/E	G/G	E/E	E/E
Aliphatic hydrocarbons (petrol, kerosene)	N	N	N	N	N	N	F
Aromatic hydrocarbons (benzol, toluol)	N	N	N	N	N	N	G
Halogenated hydrocarbons (degreaser solvents)	N	N	N	N	N	N	P
Alcohol	E – G	G	G	G	G	G	E
Animal and vegetable oils	G – F	G – F	G – F	E – G	G – F	G	G

Code: E, excellent; G, good; F, fair; P, poor; N, not recommended.
Reproduced by courtesy of G.E. Russell from a paper presented at the I.E. Aust. symposium on 'Elastomers in Engineering' held at Brisbane in October 1976.

	Common name and chemical composition of base elastomer						
	Nitrile (acrylonitrile – butadiene)	Thiokol (polysulfide)	Polyurethane (polyurethane diisocyanate)	Silicone (polysiloxane)	Hypalon (chlorusulfonated polyethylene)	Acrylic (polyacrylic)	Fluoro elastomers (fluorinated hydrocarbons)
Physical properties							
Durometer hardness range (Shore A)	10 – 95	20 – 80	50 – 90	30 – 90	50 – 95	50 – 90	60 – 90
Tensile strength (MPa)	28	10	36	10	20	16	20
Maximum elongation (%)	550	300	650	700	500	400	300
Compression set	G	P	G	G	G – F	G – F	E – G
Resilience	G	F	G	P	F	G	F
Gas impermeability	G	G	G	F	E	G	E
Electrical resistivity (polymer)	F – P	F	G	E	G	P	G
High temperature service (max. °C)	125	125	125	300	125	175	325
Low temperature service (Min. °C)	– 50	– 45	– 55	– 105	– 40	– 30	– 45
Mechanical properties							
Impact strength	F	P	E – G	F	G	P	G
Abrasion resistance	E	P	E	F	E	F	G
Tear resistance	G	P	E	F	G	P	G
Cut growth	G	P	E	F	G	F	G
Bonding to rigid materials	E – G	G – F	G	G	E	F	G
Resistance ratings							
Weather—sunlight ageing	P	G	E	E	E	E	E
Oxidation	G	G	G	E	E	E	E
Ozone cracking	F	G	E	E	E	G	E
Radiation	G	F	G	G	G	G	G
Water	E	F	G	E	G	P	E
Alkali—dilute/concentrated	G/G	G/G	F/P	E/E	E/E	F/F	E/E
Acid—dilute/concentrated	G/G	F/F	F/P	G/F	E/E	F/F	G/F
Aliphatic hydrocarbons (petrol, kerosene)	E	E	G	F	F	E	E
Aromatic hydrocarbons (benzol, toluol)	E – G	E	P	N	G	G	E
Halogenated hydrocarbons (degreaser solvents)	G – F	G	P	N	P	F	E
Alcohol	G – F	G	G	G	E	P	G
Animal and vegetable oils	G	G	E	E	G	G	E

usually by emulsion polymerization with the acrylonitrile component varying from 20 – 50% by weight. 'Hot' and 'cold' polymerization processes are employed, referring to the process temperature. 'Cold' NBR has fewer branched polymer molecules.

Properties

The most important property of nitrile rubbers is their excellent resistance to fuels and oils, even at temperatures of 140 °C. The greater the acrylonitrile content, the greater is the resistance to oils. Nitrile rubbers are processed similarly to natural rubber. They can be vulcanized with either conventional sulfur systems or peroxides. Reinforcement with carbon black improves tensile strength which can be equivalent to, or slightly better than, natural rubber. Nitrile rubbers display good ageing characteristics and abrasion resistance while their heat resistance, impermeability to gas, and resistance to many chemicals and solvents are important in some applications.

Another outstanding feature of nitrile rubbers is their compatibility with many resins, which permits easy blending. Nitrile rubber – phenolic resin combinations provide high strength adhesives for applications such as aircraft structural bonding and bonding brake linings to brake shoes. Combinations with PVC improve ozone resistance in applications such as cable insulation.

Applications

The high acrylonitrile – butadiene copolymers are used where maximum resistance to aromatic fuels, oils, and solvents is required. Applications include carburettor diaphragms, self-sealing fuel tanks, gaskets, and fuel hoses. The low acrylonitrile grades find use in areas where low temperature flexibility is important while oil resistance is secondary. Medium grades are the most widely used, typically for belting covers, shoe soles, printing rolls, rollers, sealing strips, aerosol can nozzles, hose, valves, grinding wheels, washing machine parts, and marine bearings.

Butadiene – Styrene

Butadiene – styrene or SBR is the workhorse of the synthetic rubber industry. It is commercially the most important of all rubbers—including natural rubber—because it is cheap and there is not enough natural rubber produced to satisfy world demand.

Like the nitrile rubbers, both 'hot' and 'cold' copolymers are produced. Another form the SBR is the oil extended type. Oil extension is the process of adding oil to the mix in amounts up to 35% before it is vulcanized. The action of oil on rubbers is not destructive in a chemical sense, rather oil swells the rubber and consequently softens it. Properties of oil extended, cold polymerized SBR are improved by the softening action combined with other additives. More than 50% of all SBR produced is in an oil extended form.

Properties

SBR is classed generally as the synthetic substitute for natural rubber as it displays similar properties. SBR compounds display good wear and weather resistance, excellent low temperature characteristics, low gas permeability, and good tensile properties. SBR has poor resilience, large heat build-up, low resistance to fuels and oils, and poor fatigue resistance; however, these properties may be improved by additives. The dominant role of SBR stems from economic factors related to its oil extended forms.

Applications

Oil extended SBR has major application in tire treads because of excellent wet-road traction and wear resistance. In large tires, however, SBR has been unable to repalce natural rubber because of its low resilience and greater heat build-up. Other uses of this general purpose rubber include conveyor belting, footwear, sound and shock absorption components, air seals, power transmission belts, hose, and printing press platens. SBR latexes find use in paper making, carpet manufacture, foam sponges, and paints because of their ability to hold tint. A novel use of SBR latex is to stabilize soil against erosion.

Butyl

Butyl rubber is produced from isobutylene and isoprene monomers mixed together with an inert solvent, and polymerization takes place very rapidly. The polymer separates as a precipitate and the resulting slurry is passed into hot water to flush off residual solvent and untreated monomer. The product is then filtered, hot extruded, milled, sheeted, and cooled before packing.

Properties

The most important property of butyl rubber is its extreme impermeability to gases, leading to its almost exclusive use in automotive inner tubes. Butyl rubber also exhibits outstanding chemical inertness and has exceptionally low resilience. Its energy absorption capability is high, therefore it is used in some types of equipment mounting. The rubber is serviceable over a wide temperature range and shows good resistance to heat, light, and ozone, while its electrical resistance and tear resistance—even at elevated temperature – are other advantages. Unlike NBR and SBR rubbers, reinforcement with carbon black does not improve tensile strength; however, channel and furnace blacks can improve stiffness and tear resistance.

Applications

About 75% of all butyl produced is used in the manufacture of inner tubes

125

for tires and the inner lining of tubeless tires because of its extremely low permeability. In addition, butyl has applications for steam hoses, diaphragms, flexible electrical insulation, and shock and vibration absorption. It is also used as a fabric coating for waterproofing articles exposed to weather and has had limited use as a rocket solid fuel.

References

ALLEN, P.W., *Natural Rubber and the Synthetics*, Crosby Lockwood, London, 1972

ROFF, W.J. and SCOTT, J.R., *Fibers, Films, Plastics and Rubbers*, Butterworth, London, 1971

MEARES, P., *Polymers: Structure and Bulk Properties*, Van Nostrand, London, 1965

SAUNDERS, K.J., *Organic Polymer Science*, Chapman and Hall, London, 1973

Chloroprene

Chloroprene or neoprene is produced from the reaction of acetylene and hydrogen chloride to give the monomer, which is a colorless liquid. The raw materials are corrosive, dangerous, and expensive, hence chloroprene is a fairly expensive material. It can polymerized to form a variety of products from soft, rubber-like plastics to hard inelastic polymers. It can be manufactured to resemble rubber in both its characteristics and chemical structure. Neoprene can be vulcanized by heat alone, without the use of sulfur. However, its general properties are enhanced by compounding with zinc oxide and magnesium oxide which act as vulcanizing agents.

Properties

The tensile strength of chloroprene is independent of the presence of fillers, and its hardness and creep characteristics are the same as those of natural rubber. Its elasticity is also good. The most outstanding characteristic of chloroprene is its resistance to deterioration. It has good resistance to oils and various chemical attacks. Its resistance to ozone is excellent and its light resistance is outstanding, although it tends to yellow in sunlight. As a vulcanized polymer, chloroprene is relatively unaffected by aliphatic hydrocarbons, alcohols, lubricating oils, commercial fuels and greases, hydraulic fluids (except ester and chlorinated diphenyl types), vegetable and animal oils and fats, inorganic acids, bases and salts, water and ozone. Chloroprene is severely attacked by concentrated sulfuric acid, chlorosulfonic, chromic and nitric acids, pickling solutions (especially nitric – hydrofluoric acid solutions), and by sodium hypochlorite. Its abrasion, tear, and crack resistance are good and its wear resistance is very good, especially when compounded with carbon black. Its flame resistance is good and it is self-extinguishing. The low temperature properties are fair and its gas permeability is good.

126

Applications

Chloroprene is widely used in the chemical and petroleum industries because of its outstanding oil and chemical resistance. It is used for oil and gasoline hoses and gaskets, seals, O-rings, diaphragms, and chemical tank linings. Its wear resistance makes it suitable for tires, shoes heels, etc. It is an ideal material where electrical insulation must be in contact with oil. Its ozone resistance makes it useful in electrical systems where there is arcing or electric discharge. It is also used in heavy duty engineering components such as bridge bearings, where it competes with natural rubber. Its self-extinguishing characteristics make it very suitable for coal mine conveyor belts.

Other applications include wire and cable coatings and floor tiles. Chloroprene can be extruded or molded and is available as sheets and hoses. Chloroprene can be identified, although it is nonflammable, from the characteristic that it burns with a sooty flame if kept hot and leaves a swollen black residue. It can be distinguished from PVC by decomposition in concentrated nitric acid, by which PVC is virtually unaffected.

Chlorosulfonated Polyethylene

Chlorosulfonated polyethylene was first introduced in 1952 and has the trade name 'Hypalon'. It is produced by reacting polyethylene with chloride and sulfur dioxide. The process is complex and, as with chloroprene, the raw materials are dangerous, which in part accounts for its fairly high cost.

Properties

Although the tensile strength and elasticity are not as good as chloroprene, Hypalon has better elongation. Its tensile strength is not affected by fillers. Hypalon can be vulcanized without additives and it has the highest colorability of all synthetic rubbers. Inclusion of chlorine in the polymer results in high flame resistance and it is self-extinguishing. Even a thin coating can impart a high flame resistance to flammable materials. Electrical properties make it appropriate for low voltage insulations. Long-term weathering, water immersion, and underground exposures do not affect electrical characteristics.

Properly compounded, Hypalon is unexcelled in its resistance to ozone because of its saturated bonding. Ozone resistance is so complete that it is used for gaskets in commercial ozone generators. Ageing characteristics and compression recovery characteristics are good, as are low temperature properties. Abrasion, wear, fatigue, impact, cutting, and cracking resistance are also good. The elastomer is resistant to ultraviolet radiation deterioration. It has high resistance to damage by oxidation, water, oil, and weathering. It can be lastingly colored without deterioration, indoors or outdoors. The elastomer is highly resistant to chemicals, oils, grease, and fluids, particularly to oxidizing agents such as concentrated sulfuric acid, hypochlorite solutions, and to strong and weak alkalis. Like other chlorine-containing polymers, it is immune to bacteria and micro-organism attack.

127

Applications

Hypalon was used initially in pump and tank linings, tubings, and other areas where chemical resistance was required. It is now used also in applications where its weatherability, colorability, heat, ozone, and abrasion resistance as well as electrical properties are important. Flexible hose in oil and chemical applications, tank linings, control cable for atomic reactors, automotive ignition wiring, and cable insulation are other applications.

Heavy duty uses include conveyor belts for high temperature applications. It is used as V-belts, motor mounts, O-rings, seals, and gaskets, and for shoe soles and garden hoses. Hypalon can be extruded as a protective and decorative veneer as sealing and gluing strips. It is often used as a protective coating for neoprene and natural rubber components. Hypalon can be identified by means of a dilute solution of chlorosulfonated polyethylene in pyridine, which gives an orange or red color when treated with fluoroamine.

Epichlorohydrin

Epichlorohydrin is a common monoepoxy used in the manufacture of resins. In the elastomeric form it is available as a homopolymer and as a copolymer of epichlorohydrin and ethylene oxide, the latter usually containing about 40% ethylene oxide. Manufacture of epichlorohydrin is by chlorination of acridine, or high temperature chlorination of propylene, or directly from allyl chloride by oxidation with peracetic acid. Both homopolymers and copolymers are saturated special rubbers curable with diamines or ammonium compounds.

Properties

When reinforced, these high priced rubbers have average tensile strength and elongation properties and an unusual combination of other characteristics including low heat build-up. The homopolymer has outstanding resistance to ozone and organic liquids and has good resistance to swelling due to hot oil. It also has extremely low permeability to gases and excellent weathering properties. The homopolymer with a chlorine content of about 38% is flame resistant and has low resilience characteristics and low temperature flexibility (to -15 °C). The copolymer is more resilient and has low temperature flexibility to -40 °C but it has poorer permeability characteristics. The copolymer rubber has good resistance to ozone, and good resistance to swelling by oils.

Applications

Epichlorohydrin rubbers are excellent for use where repeated shock or vibration exists, such as in equipment mounts and vibration dampers. Other applications are in the manufacture of diaphragms, seals, and gaskets, hoses for petroleum handling, and low temperature parts. Epichlorohydrin is an expensive rubber. Quantities produced are small in comparison with large tonnage rubbers. Its trade name is 'Hydrin'.

Ethylene – Propylene

Ethylene – propylene rubber is produced by the copolymerization of ethylene and propylene using catalysts such as titanium and vanadium. Commercial products consists of two types:

1. saturated copolymers, designated as EPM, which can be vulcanized only with special techniques;
2. terpolymers, designated as EPDM, which contain a small amount of diene to enable vulcanization with sulfur.

The copolymer type is an exceptionally interesting rubber. The two monomers are cheap and plentiful, but the polymerization process is not. The rubber is exceptionally resistant to oxygen, ozone, and heat. Molecular structure of the copolymer is similar to natural rubber but without the double bonds which are responsible for the ageing deficiencies of unsaturated rubbers.

Properties

Ethylene – propylene rubbers are light gray to amber colored. They have hardly any odor and have good ageing characteristics. EPDM has some outstanding properties. It has exceptional resistance to heat, oxygen, and ozone and is easily vulcanized and processed. Its tack properties are below average. The specific gravity of EPDM is lower than that of other major rubbers. Since rubber products are sold by volume rather than by weight, this can be an important cost consideration. EPM and EPDM may be oil extended, which also reduces cost.

Applications

Ethylene – propylene rubber has excellent electrical properties and high ozone resistance, making it ideal for cable coverings. Other uses include automobile interior trim, footwear, tires, conveyor belts, hose, seals, and gaskets. Most of the annual production of EPDM is used by the automobile industry.

References

LEE, H.L. and NEVILLE, K., *Handbook of Epoxy Resins*, McGraw-Hill, New York, 1967
ROFF, W.J. and SCOTT, J.R. *Fibers, Films, Plastics and Rubbers*, Butterworth, London, 1971
Chemistry in the Economy, American Chemical Society, Washington, DC, 1973
ALLEN, P.W. *Natural Rubber and the Synthetics*, Crosby Lockwood, London, 1972

Fluorosilicones

Fluorosilicones are silicone rubbers (FSR) and have the characteristics of the general purpose silicones: resilience over a wide temperature range, resistance to ozone and weathering, and outstanding electrical properties. Additionally, fluorosilicone rubbers withstand oils, fuels, and solvents at high and low temperatures. They show superior corrosion resistance and are available as rubbers and adhesives.

Properties

Fluorosilicones have fair-to-good resilience, good hystersis resistance, and high elongation. Flexural cracking resistance is good at both slow and fast loading rates. Fluorosilicones have good damping properties and tear strengths. Unlike many elastomers, FSR has poor abrasion resistance. They have good thermal properties with a serviceable temperature range of 30 – 200 °C for continued use. Low temperature stiffening occurs below – 70 °C. FSR resists compression set and has good elastomeric properties at high and low temperatures. Tensile strength does not vary as much compared to other rubbers. Flame resistance is poor. FSR gradually hardens and loses elasticity when aged at temperatures from 150 to 315 °C; however, resistance to compression set and attack by oils and chemicals improves with heat ageing.

FSR possesses design and processing versatility. Fabrication methods include molding, extruding, calendering, dispersion coating, and sponging. Parts can be made in a range of shapes and sizes by compression, transfer, and injection molding. Tubing, rods, seals, and wire insulation can be made by extrusion. FSR dispersed in a solvent is applied to fabrics made of glass, nylon, Dacron, and other fibers to improve their flex life and impart insulating properties. Sponge rubbers are excellent for damping, cushioning, and protecting components exposed to harsh environments.

Uses

Because FSR is expensive compared to most other elastomers, its uses are restricted to parts requiring resistance to oils and solvents at high temperatures, e.g., seals, gaskets, O-rings, and industrial hoses. Fluorosilicone adhesives are used for fuel and solvent-resistant bonding in chemical processing and transportation industries. They maintain excellent adhesion when exposed to fuels, oils, and solvents.

Hard Rubber

Hard rubber was produced originally from natural rubber and was called 'Ebonite'. The bulk of hard rubber today is made with styrene – butadiene rubber and is commonly called SBR. Once it has been formed, hard rubber cannot be returned to its original state, being a thermosetting plastic. It differs, however, from other themosets such as the phenolics and ureas in that it will resoften somewhat under heat. This characteristic resembles the thermoplastic acetates, polystyrenes, and vinyls. To be classified as hard rubber, sulfur must be present in the compound in amounts greater than 18% and up to a saturation point of 47%, at which the material is vulcanized completely.

Properties

Important properties of hard rubber are the combination of relatively high tensile strength, low elongation, and extremely low water absorption. When reinforced with carbon black, the mechanical properties are good. It has very good abrasion resistance, but is slightly less resilient than natural rubber. Swelling and adhesion properties are similar to those of nitriles and its ageing resistance is slightly superior. It has a service temperature range from -60 to 80 °C for most grades.

Hard rubber may be compression, transfer, or injection molded. In sheet form it can be hand fabricated into many shapes. Its machining qualities are comparable to brass and it may be drilled and tapped. The material lends itself readily to permanent or temporary sealing with hot or cold cements and sealing compounds. Size and shape of hard rubber parts are dependent only upon press equipment size and vulcanizing.

Uses

The largest application for hard rubber is in battery boxes. The water meter industry is also a large user. Hard rubber linings and coatings, either molded or layed-up, account for large quantities of material. In the electrical industry, hard rubber is used for terminal blocks, insulating materials, and connection protectors. The chemical electroplating and photographic industries use large quantities for acid handling devices. Sports equipment, adhesives, and dentures are also produced from hard rubber.

Natural Rubber

Natural rubber is a material made from natural latex, a white liquid resembling milk tapped from rubber trees, the chief one being *Hevea brazilialas*. The latex is dried by adding acetic acid which coagulates it. The product is then rolled into sheets and baled. Most rubber is produced in South East Asia with small amounts in South America and Africa.

The main vulcanizing agent for natural rubber is sulfur. Another important additive is carbon black, which strongly reinforces the rubber and increases tensile strength. Carbon black also increases modulus of elasticity, hardness, abrasion resistance, and electrical conductivity, but decreases elongation.

131

Properties

Rubber is highly deformable and will undergo essentially full recovery. Temperature has a marked effect on the tensile properties of natural rubber. As temperature decreases, strength increases and elongation decreases until the rubber becomes brittle. Natural rubbers have a higher tear resistance than synthetic rubbers, especially at high temperatures.

The ability of rubber to be vulcanized makes it susceptible to oxidation. Although antioxidants are added, the oxidation resistance of natural rubber is not as high as that of most synthetic rubbers. Rubber, especially in a stretched condition, is attacked by ozone. Natural rubber is inferior to synthetics in oil and solvent resistance. Oils tend to swell it and cause deterioration. The material ceases to be 'rubbery' below $-73\ °C$ and may lose all stiffness at about 70 $°C$ depending on sulfur content.

Applications

Gum rubbers contain $80-96\%$ natural rubber. Typical products are balloons, rubber bands, and transparent rubber articles. Applications with $50-80\%$ natural rubber content include sponge rubber, tire treads, inner tubes, footwear, conveyor belts, and seals. Hose, tubing, shoes, and electrical insulation are uses with $30-50\%$ natural rubber content. Compounds containing $10-30\%$ natural rubber are used for brake linings, mats, and floor tiles. Other applications include rubber springs, energy absorbers, vibration isolators, and noise suppression materials.

Plastic and Rubber Foams

Plastic Foams

Plastic foams are two-phase systems of a gas dispersed in solid plastic. They are known alternatively as cellular, foamed, or expanded plastics, or as plastic sponges. Plastic foams may be thermosetting or thermoplastic materials, the type and properties of which are governed by the pressure of gas inside the voids or cells, the surface tension forces which cause plastic flow, and the viscoelastic forces which restrict plastic flow.

The thermal conductivity of foams is related to the amount of gases enclosed. There is a small dependence of thermal conductivity on bulk density. Plastic foams are efficient in reducing reverberation due to reflection of sound waves. The chemical properties of a plastic expanded into a foam do not generally change; however, the stresses induced during expansion make some plastics more susceptible to solvent attack. The majority of plastic foams are either opaque or translucent. Foam colors can be varied by addition of dyes or pigments to the plastic prior to expansion. Many foams are resistant to fungus and bacteria.

132

Methods of plastic manufacture include use of expandable beads or pellets, reaction foaming, froth foaming, spray application, and injection molding.

Thermal insulation is the largest application of rigid plastic foams. Typical examples are refrigerators, sandwich panel covers for building, and insulation in the transportation field. Plastic foams are also used for acoustic insulation and buoyancy. They are used in the packaging field because of their insulation, cushioning ability and cost advantages.

Rubber Foams

Rubber foams are manufactured from natural rubber and employ air or a gas introduced into the latex to create a foam. The foam is then gelled, cast into molds, vulcanized, and dried. Advantages of rubber foams are good elastic properties (even under severe loading), cushioning, and acoustic properties. A limitation of rubber foams is that they will not generally withstand high temperatures (250 °C). Carpet backing is perhaps the most common use of latex foams. They are also used extensively for automobile seats. Usually blends of natural and synthetic rubbers are used since foams such as polyurethanes can be made much cheaper.

References

ALLEN, P.W. *Natural Rubber and the Synthetics*, Crosby Lockwood, London, 1972

GOBEL, E.F. *Rubber Springs Design*, Newnes-Butterworth, London, 1974

LEVER, A.E. and RHYS, J.A., *The Properties and Testing of Plastic Materials*, Temple Press, Middlesex, 1968

Polyacrylates

Polyacrylates are tough rubbery polymers of ethyl and butyl acrylate cross-linked with long-chain amines. Like other special purpose synthetics, they have good resistance to ageing and to oils, although they are difficult to process. They are characterized by excellent stability against oxidation and ozone, and resist severe temperatures.

Polyacrylates are processed similarly to conventional rubbers but are cured with amines which control ageing properties. Other properties can be altered by fillers such as carbon black. Polyacrylates are available in a variety of colors. Products are usually made by extrusion. However, less pressure is needed in molding the rubber than in extruding it.

Properties

Characteristics of polyacrylates include:

temperature resistance to 200 °C

resistance to oxidation and ozone at normal and high temperatures
good flexing life
resistance to fading
good resistance to swelling and deterioration in oils
good permanence of color, especially white and pastel shades
remarkable dielectric stability under severe conditions

Applications

Application of polyacrylate rubbers is mainly confined to seals and gaskets in the automotive industry, such as automatic transmission seals, extreme pressure lubricant seals, and searchlight gaskets. Other uses are in conveyor belts, printing rollers, power transmission belts, and tank linings. They are also used as protective coatings on cotton and nylon for ropes and are applied to cotton duck for belting. Additional applications are in garden hoses and pigmented binders for paper, textiles, and fibrous glass.

References

ALLEN, P.W., *Natural Rubber and the Synthetics*, Crosby Lockwood, London, 1972

DEANIN, R.O., *Polymer Structure, Properties and Applications*, Cahners, Boston, Mass., 1972

KENNEY, G.F., *Engineering Properties and Applications of Plastics*, John Wiley, New York, 1957

Polybutadiene

Polybutadiene is cheap, very resilient (more so than natural rubber), and wears well on tire treads. It is difficult to process and has some undesirable properties. It is widely used when blended with other rubbers, especially SBR and natural rubber.

Polybutadiene is made by solution polymerization. By using different catalysts, polybutadiene can be made with structure ranging from almost 100% *cis* to 100% *trans*. Materials of the latter type can be used to replace gutta perch in items such as golf ball covers. In the rubbery range, there are two levels of *cis* content: 'high-*cis*' (about 80%) and 'medium-*cis*' (about 45%).

The unsubstituted hydrocarbon chains of polybutadiene have no polarity to produce any significant intermolecular attraction, thus each chain acts flexibly. The molecules are high entangled, however, and prevent liquid flow under stress. The entanglements aid elastic retraction when external stress is removed. The molecules disentangle gradually under sustained stress, which results in low mechanical properties and high creep rates. Vulcanization produces a few crosslinks, which improves hot and cold strengths, creep resistance, and recovery. The molecular flexibility also produces high abrasion resistance.

The unsaturated carbon bonds cause sensitivity to oxygen and ozone (especially when stretched) and to oxidizing chemicals. This sensitivity is increased by heat and light, resulting in poor ageing properties. To overcome these shortcomings, stabilizers are added. One such stabilizer is carbon black, but it severely limits colorability. Poor stability of the carbon bonds usually result in poor color stability.

Properties

Some of the important properties of polybutadiene include:

higher resilience than SBR and natural rubber
difficult to process alone, hence blends are used
wear resistance in blends with SBR is good
cis-polybutadiene is flexible
temporary crystallization of polymer chains on stretching results in good
  strength

Applications

Polybutadiene is used in passenger vehicle tires. A blend of 75% SBR with 25% polybutadiene is the most common rubber is use for tires. The low resilience of SBR is used to decrease the skidding tendency of tires and polybutadiene increases wear resistance and resistance to groove cracking. Other uses are in footwear, conveyor belts, seals, floor tiles, and sponge compounds.

Polyisoprene

Cis-polyisoprene is synthetic natural rubber. It is often blended with natural rubber for economic reasons. The main feature of polyisoprene is superlative resistance to heat and atmospheric ageing. The polymer has a methyl side group on every fourth main-chain carbon atom, which reduces molecular flexibility and results in low resilience and low temperature flexibility. *Cis*-polyisoprene does not contain aldehyde groups in its molecules and hence does not 'storage harden' as natural rubber does.

Properties

Polyisoprene has superior resistance to heat and ageing. It does not fade in sunlight and does not storage harden. Polyisoprene has high tear strength, second only to natural rubber. Its low moisture absorption allows its use for electrical insulation. Flame resistance, however, is poor.

Applications

Polyisoprene is used in general rubber products: in tires (as blends) and for electrical insulation. Other uses include power transmission belts, footwear, and molded mechanical goods. The medium-*cis* type is used in injection molding because of its good flow behavior. Competition between natural rubber and polyisoprene is not a question of processing, end use, or properties. Competition is in terms of supply and price, polyisoprene being more costly to produce than natural rubber.

Polypropylene Oxides

Polypropylene oxide/butadiene rubbers are low-to-medium density materials. They have excellent impact and tear strengths, as with natural and synthetic rubbers. Mechanical properties and resilience are very good. They are among the better synthetics in terms of low temperature performance, although elevated temperature performance is only fair.

Polypropylene oxides possess excellent environmental resistance to oxygen, ozone, and weathering, but suffer degradation from acids, hydrocarbons, and radiation. They are used principally for electrical insulation and molded mechanical goods.

Polysulfides

Commercial forms of polysulfides include solid and liquid rubbers, prevulcanized molding powders, and latex. The solid rubbers range from brown to white in color and some types have an unpleasant mercaptan-like odor. Polysulfides were among the earliest rubbers to be discovered. Polysulfides are made by a condensation reaction of an aliphatic dihalide and sodium polysulfide, the simplest product being derived from ethylene dichlorate and sodium tetrasulfide. In commercial production, magnesium hydroxide is used to assist dispersion and facilitate the reaction. The latex so obtained may be coagulated with acid then washed, dried, and pressed into blocks.

Properties

Water absorption of polysulfides is affected by polymer type, sulfur content, and compounding. All types swell in hot water. Permeability is generally much lower than that of natural rubber and is comparable to that of butyl rubber. Polysulfides are not resistant to high temperatures.

Polysulfides burn rapidly and, despite their high sulfur content, have sufficiently high heats of combustion to be used as fuels for some rocket motors. Solid forms have limited electrical applications. Liquid types, however, are used in encapsulating electrical components. Properties of the cured end products of both types are similar. Uncured, unfilled polysulfides have very low strength and require reinforcing fillers. Their strengths and resilience, however, are not comparable to natural rubber or general purpose synthetics. Compression set and abrasion resistance are also poor.

136

Solid polysulfides can be processed on conventional machinery, although with somewhat greater difficulty than natural rubber or general purpose synthetics. Polysulfides often have an offensive odor. Their outstanding advantages are exceptional resistance to oils and solvents and low permeability to gases. Unless specially compounded, polysulfides may be attacked by micro-organisms and insects.

Applications

Whereas during the Second World War many different grades of solid poly-sulfides were used in most natural rubber applications, many of these have been discontinued and a more limited range of uses remain. These include linings of oil and paint hoses, printing rolls, molded seals, coated fabrics, and some cable coverings.

Liquid polymers are extensively used in sealants for building, civil engineering, boat building, also for cast printing plates and rollers, impregnated leather gaskets, paints and adhesives, flexible molds, and cast solid rocket fuels. Some liquid types are used as vulcanizing agents for other sulfur-cured rubbers. Latex, sometimes blended with PVC or polyvinylidene chloride, has found use as protective coatings and linings, e.g., in concrete fuel tanks. Flame-sprayed polysulfide powders have been used to form anti-cavitation coatings on propeller shaft casings and rudder struts of ships.

Styrene Butadiene Styrene and Styrene Isoprene Styrene

The products of SBS and SIS are known as block copolymers. They have characteristics similar to rubber at ordinary temperatures but may be processed at elevated temperatures in the same manner as conventional thermoplastics. These two products are often known as 'thermoplastic rubbers' because their ease of processing is similar to that of polyethylene.

SBS is composed of 15 – 30% styrene, the remaining material being buta-diene, polybutadiene, or styrene – butadiene rubber. SIS is composed of approximately 15 – 30% styrene and the remaining material is polyisoprene. The polystyrene blocks in each molecule chain associate into domains which act as crosslinks, thus there are aggregates of styrene blocks in a butadiene matrix. At ordinary temperatures, the styrene domains are rigid and immobilized at the ends of butadiene chains. The butadiene blocks create rubberiness and can be regarded as polybutadiene with polystyrene crosslinks.

Properties are controlled by molecular weights, styrene and butadiene blocks, and by the ratio of styrene and butadiene. SBS and SIS properties are comparable with those of natural rubber. They are not outstanding, but there is reasonable tear strength and abrasion resistance, fair resistance to oxygen, ozone, and weathering, and good electrical performance. They exhibit poor resilience, heat build-up, and poor flame resistance. Main applications are in adhesives, footwear, carpet backing, and as additives to plastics.

Urethanes

Urethanes are materials which range from durable oil-resistant rubbers and fibers (either relatively inextensible or extensible and elastic) to surface coatings, adhesives, and flexible or rigid foams. They are not direct polymers of urethanes but are derived from the reaction of polyesters or polyethers with di- or polyisocyanates to produce complex structures containing urethane linkages. They are special purpose elastomers with properties generally between rubbers and thermoplastics. The resulting product depends on the ingredients, degree of crosslinking, and method of manufacture. All urethanes are expensive but they are strong and offer exceptional resistance to wear and oils.

Flexible Foam

Flexible foams are resilient open-cell structures and are by far the most widely used form of polyurethane. Foams are produced using a 'one-shot' process. The basic ingredients and catalysts are mixed immediately before foaming is required. The reacting mix is then allowed to foam on to a slowly moving trough or into molds. The reaction is exothermic and the heat produced assists curing. The foam has good flow properties before curing.

The properties of flexible foam are similar to those of natural rubber and SBR foams, but polyurethanes are less flammable and have better resistance to ageing, oxygen, and ozone. They also have lower density. The main uses for flexible foam are cushioning, upholstery, and safety padding. Other uses are paint rollers, sponges, and packaging for delicate equipment.

Rigid Foam

Rigid foam is the second largest use of polyurethane. A 'one-shot' process is used similar to that for flexible foams; however, more effective crosslinking is employed to give a rigid structure. Such foams have a particular advantage in that they can be formed *in situ*, for example, in wall cavities for thermal and acoustic insulation. Polyurethanes act as adhesives to most cavities, using the surrounds as a protective skin. Foams have an economic advantage over polystyrene because of cheaper starting materials and lower density. Polyurethanes have better oil, grease, and heat resistance than polystyrenes and have lower thermal conductivity. Main uses for rigid foam are in refrigerated railroad cars, household and commercial coolers, and structural and insulation sandwich panels in roofs, ceilings, aircraft, and marine buoyancy applications.

Elastomers

All synthetic rubbers have their own particular advantages because of some limitation of natural rubber, whether it be a property disadvantage or a supply/cost advantage. In general, their elasticity is due to highly flexible segments,

138

low internal forces, and little or no crystallinity. Polyurethanes can be produced with a number of desirable properties such as creep resistance, high tensile strength, tear resistance, and abrasion resistance. Polyurethanes have some outstanding properties: higher tensile strengths than other rubbers, excellent tear and abrasion resistances, outstanding resistance to oxygen and ozone, and good hardness. Polyurethanes, however, exhibit low resilience and decomposition under prolonged exposure to water or steam. Main uses of the elastomers are in oil seals, solid rubber industrial tires, shoes, diaphragms, and industrial chute linings.

Fibers

The polyurethane rubber thread known as 'Spandex' has the capability of resisting numerous machine washings and drying and is therefore used extensively in foundation garments, swimwear, and surgical hose. Another fiber is that known as 'Pirlon U', its main competition being some nylons. Advantages over nylons are higher tensile strength, less propensity to discoloration in air, and lower moisture absorption. The advantages of nylon over Pirlon U are its ease of handling and its being less wiry and harsh. Polyurethane fibers are used mainly in high humidity atmospheres for electrical insulation; although nylon is cheaper, it is not as effective.

Adhesives

Many isocyanates have good adhesive properties and have been particularly successful for the bonding of rubbers. Polyurethane products have been glued to tougher products such as to metals and fabrics in the shoe industry.

Coatings

Coatings based on polyurethanes have good abrasion and impact resistance and are flexible. They can be dipped, sprayed, or brushed on. Examples are coatings for wood finishes in marine applications, rain erosion protective coatings for aircraft and compression blades, and fabric coatings. Polyurethanes resist degradation by micro-organisms and mildew. They can be sterilized and with care can be dry cleaned. They are resistant to gasoline, oil, and ozone and exhibit good tear and abrasion resistance. Toxic gases from their combustion limit or preclude some applications.

Identification of Plastics and Rubbers

The recommended approach to the identification of a given unknown plastic or rubber sample is as follows:

preliminary examination

initial tests
heating tests
elemental analysis
final identification

Common sense and experience may indicate a better order than above in some cases.

Preliminary Identification   First, determination must be made of whether the material is a rubber, a flexible thermoplastic, a rigid thermoplastic, or a thermosetting plastic. Appearance, method of fabrication, rigidity, and the effect of heat are assessed. After a few basic tests, the type of material should be known.

Initial Tests   To aid further identification, several tests may be performed:

1. *Beilstein test*. Identifies halogens and is performed by heating a piece of copper wire to red heat and rubbing it on the sample—if a green flame is noted when the copper wire is reheated, halogens are present.
2. *Specific gravity*. A beaker of water is required into which the sample is placed and noted as being either lighter or heavier than water.
3. *Bounce test*. Most rubbers bounce but butyl rubber does not. When moldings from toughened polystyrene are dropped they give a peculiar metallic ring.
4. *Odor test*. Some materials have pronounced odors, e.g., polysulfides.
5. *Feel*. Polyethylene and PTFE have a waxy feel unlike other polymers.
6. *Color*. Whether the material is black, colored, or transparent can give a lead to its identification.

Heating tests   Only small quantities should be used as some materials may explode or evolve toxic gases. Color of flame, flammability, and self-extinguishing or self-supporting combustion should be noted. Reference information will usually identify the material.

Elemental Analysis   Elements such as nitrogen, phosphorous, and sulfur may require chemical analysis.

Final Identification   Mass spectrography or chemical analysis with acids and reactants provides a final means of identification.

It is essential that results of all tests performed should be recorded as they are obtained, even if their meaning is not immediately apparent. In this manner all evidence can be considered and results then usually fall into place.

References

SAUNDERS, K.J. *Identification of Plastics and Rubbers*, Chapman and Hall, London, 1967
Aids to the Identification of Plastics—Explanatory Notes with Samples, Building Research Establishment, UK Department of Environment, January 1977

9
Other Engineering Materials

Carbides

Carbides are characterized by very high melting points and high hardness. There are basically two types: refractory carbides and cemented carbides. The most important refractory types are the carbides of silicon and boron, other less important ones being carbides of the transition elements. Cemented types are the carbides of tungsten, titanium, and tantalum embedded in a binder metal, usually cobalt.

Of the refractory carbides, silicon carbide or carborundum is the oldest. It is produced in an electric furnance from sand and coke. The product is then ground to the required size and mixed with a bonding agent. Silicon carbide is characterized by high thermal conductivity, low thermal expansion, good abrasion resistance, high strength, and high cost. Silicon carbide crystals are used as semiconductors and as whisker reinforcement for plastics to produce materials with very high strengths and stiffnesses. Sizes and shapes of such materials, however, are limited by fabrication difficulties.

Boron carbide also has high hardness and wear resistance and is an excellent neutron absorber. Its melting point is 2450 °C, but use is restricted to 980 °C as it reacts with oxygen at higher temperatures. Because of its excellent neutron absorbing ability, boron carbide is used in power reactors. Other uses are similar to those of silicon carbide such as abrasives, wear resistant products, and parts requiring high heat resistance.

Tungsten carbide is a cemented carbide having very high hardness (Moh 9.5) and consequently has its main application in the tooling industry. Tungsten carbide dies can only be ground with diamond impregnated wheels and drills. Other uses include hard facings and erosion resistant parts.

Titanium carbide is used for cutting tools and heat resistant parts. It is less costly than tungsten carbide but tends to be more brittle. Combinations of titanium carbide and tungsten carbide are used in a cobalt matrix to provide a relatively inexpensive cutting material.

An alloy of tungsten, cobalt, and vanadium carbide has been developed recently for high speed cutting tools. Molybdenum and zirconium carbides also find use in high temperature tool steels. Some carbides have poor thermal shock resistance.

Clad and Plated Metals

There is a great variety of clad and plated metal combinations that provide properties superior to other metals at lower costs or provide properties not found in any metal alone. Cost reduction results from the use of a relatively thin film of protective expensive metal which is bonded to a cheaper base metal that provides strength. Clad metals differ from plated metals in the relative thickness of the cladding or plating material used. Clad metals are usually employed where the environment is severe and additional impervious thickness is warranted. In general, plating thicknesses are only a few tens of nanometres, whereas cladding thicknesses may be measured in millimetres.

Clad Metals

In the case of steels, cladding is the strong permanent bonding of high alloy steel or other hard materials to plain carbon or low alloy steel. Clad materials may provide the service equivalent of homogeneous alloy steels with cost savings up to 65%. Clad steels were first used to supplement the use of nickel and allowed development of large corrosion resistant parts. Carbon or alloy steels clad with stainless steel constitute the major tonnage, but cladding with aluminum, copper, monel, nickel and titanium are also common. Fabrication processes usually employed are:

> pack rolling
> plate cladding
> explosive cladding
> extrusion
> continuous brazing
> casting
> welding

Several of the above less familar processes are outlined next.

Pack Rolling   A pack or sandwich of constituent materials is heated to a pre-determined temperature (usually red heat) and rolled to the desired gage in pack rolling. It is then trimmed to produce clad plate of uniform thickness. Bonding takes place during the heating process.

Plate Cladding   A slab of carbon steel and a plate of nickel are heated and rolled, bonding taking place during rolling, to form plate cladding.

Explosive Cladding   A slab of copper, for example, may be sandwiched between layers of cupronickel alloys, with separations only due to surface irregularities of two mating surfaces. An explosive is applied to each side and detonated in explosive cladding. The bonded material is then rolled. The process is used in producing silverless coins.

Continuous Brazing   Brazing alloy is is distributed between two strips which are continuously heated, fused, and rolled in continuous brazing. Quenching and tempering may be employed to enhance properties.

142

Applications Clad materials can be cut, machined, rolled, and welded by conventional means. However, since dissimilar metals can have large differences in expansion coefficients, thermal effects may occur. Clad metals are available in a variety of forms: strip, sheet, foil, plate, wire, rod, and tubing, and are used in applications such as bridges, nuclear power reactors, and corrosion resistant platings. The electrical industry uses many clad materials. Silver bonded to copper is used in high temperature coils and high frequency conductors. Aluminum bonded to copper provides a useful magnetic wire, and silver bonded to nickel results in good wear resistance of electrical contacts. Gold bonded to copper is used in chemical process equipment. Metal cladding is usually employed for continuous forms or for large surface area. For smaller complex shapes, the plating process has many advantages.

Plated Metals

Before metals can be plated, the surfaces must be prepared properly for good adhesion. Cleaning techniques include mechanical (such as blasting or tumbling), chemical (such as alkaline, acid, or organic agents), and electrolytic methods.

Electroplating Used widely in applying decorative and protective coatings to metals. Most metals can be electroplated, but those commonly plated are nickel, chromium, cadmium, copper, silver, zinc, gold, and tin. In commercial plating the object to be plated is placed in a tank containing an electrolyte. The anode consists of a plate of pure metal and the object to be plated is the cathode. The tank contains a solution of salts of the metal to be applied. Low voltage d.c. current is usually required. A surface layer of pure metal is thus deposited on the cathode.

Chrome Plating For wear and abrasion resistance, chromium is used. Coatings are seldom less than 0.05 mm thick.

Galvanizing A zinc coating used extensively for protecting low carbon steel from atmospheric corrosion. It offers a low cost coating with good appearance and good wearing properties. Zinc baths at temperatures of about 450 °C are used as dip tanks. Continuous and automatic dipping processes may be used for sheet and wire coatings. Zinc coatings may also be applied by spraying molten zinc on steel. Sheradizing is produced by tumbling parts at elevated temperature in zinc dust. Zinc may also be applied by electroplating.

Tin Coatings Applied to sheet steel for use in food containers, accounting for the use of approximately 90% of tin produced. These are applied by electro-tinning or hot dipping. Because of porosity with plated tin coatings, lacquer seals are often used to provide additional proctection.

Other Plating Materials Copper is often used as an undercoat for nickel plating as it provides good adhesion. Nickel plating is popular for protecting steel or brass from corrosion and for presenting a bright appearance. Lead has only limited commercial use as a coating against certain acids. For nonferrous articles used in food handling, silver plating is widely used. Many plating materials are applied by metal spraying, which includes metallizing, metal powder spraying, and plasma flame spraying.

Coatings to Improve Wear Resistance

Various coatings and coating techniques have been developed to combat deterioration of metal surfaces due to wear. Wear can result from either of two interactions: adhesion by cold welding of surfaces at their points of contact or by abrasion, i.e., the ploughing of a hard surface into a softer one. The basic requirement for a coating is to reduce the adhesive and abrasive interactions.

Although oils and greases possess adequate wear reducing properties when subjected to many environmental conditions, such lubricants are limited for certain applications due to:

high temperature
chemical degradation when exposed to high temperature
change in viscosity with a change in temperature
lack of lubricating film under fretting conditions
relubrication required in inaccessible areas
operating under high stress conditions

Promising coating systems are those based on laminar solids such as molybdenum disulfide and graphite bonded to the surface by resin or an inorganic matrix. Ceramics also have been investigated for antiwear applications. Refractory materials such as alumina, chromium oxide, iron boride, and molybdenum oxide tend not to cold weld during sliding contact and can withstand high operating temperatures. They are characterized by high hardness and chemical stability. Ceramics may be deposited by plasma arc spraying.

Soft metals such as silver, gold, lead, and indium can be used in thin film over hard materials to provide lubrication and to prevent cold welding. Properties such as low shear strength, good bonding, high thermal stability, high heat capacity, and high thermal conductivity are required for solid lubrication. Conductivity is advantageous when it is necessary to pass an electrical current through a moving part. Plastics can also be bonded to metal to provide nonconducting, moderately low temperature, wear resistant coatings.

Honeycomb Materials

Honeycomb panels may be compared to I-beams. The facings correspond to the flanges, having high strength/high stiffness material as far from the neutral axis as possible and increasing the section modulus. The honeycomb core is comparable to the web and carries shear stresses. However, honeycombs differ from I-beams in that they maintain continuous support for the facings. The facings can therefore be worked up to or above their yield strength without

144

buckling depending on the core cell size. The adhesive which bonds the honeycomb to its facings must be capable of transmitting the shear loads between core and facings.

When honeycomb sandwich panel is loaded as a beam, the honeycomb and bond resist the shear loads while the facings resist the bending moments. When loaded as a column, the facings alone resist the axial forces and the core stabilizes the faces against buckling.

Honeycomb sandwiches are more efficient than I-beams. The combination of high density facings and low density cores provides much higher section modulus per unit density than any other known construction method, particularly for highly loaded structures. For equivalent rigidity, the weight of aluminum faced honeycomb structure is one-fifth that of hardwood plywood, one-tenth that of solid aluminum, and one-sixteenth that of steel.

Facing Materials

Any thin bondable material with high tensile or compressive strength-to-weight ratio is a potential facing material for honeycomb panels. Some materials used are as follows:

Aluminum The high strength aluminum alloys are commonly used as facing materials for structural and nonstructural applications. Where temperatures exceed 200 °C for long periods, they are generally not used.

Stainless Steel and Superalloys Because of their good strengths at elevated temperatures, stainless steel and superalloy facings are used in aircraft. Some perform satisfactorily at temperatures of 1100 °C. Porcelain enameled steel is another facing material.

Glass Reinforced Plastics Where parts have a complex contour, glass reinforced plastics are ideal. They offer special advantages as follows:

> excellent weight control
> easy build-up of material where it is needed and thin faces are permissible
> where excess material is not needed
> excellent heat insulation
> excellent-to-medium temperatures strength
> high strength-to-weight ratio
> high impact strength
> compatibility with other materials

Plywood Plywood and veneer facings are used for structural and nonstructural applications. The most common use of plywood facings is in interior and exterior doors of buildings. Plywood facings for doors are usually bonded to paper honeycomb cores.

Resin Impregnated Paper Paper facing materials are used generally for flat nonstructural panels such as interior walls and partitions.

Core Materials

Any material that can be made into a foil or thin sheet, welded, brazed, or adhesively bonded can be made into a honeycomb.

Aluminum   Cores are supplied in expanded or unexpanded forms. Thicknesses vary from 1.5 to 45 mm or more. High strength aluminum alloy cores are available in densities ranging from about 20 to 130 kg/m³.

Stainless Steel   Stainless alloys are generally used where temperatures are too high to permit use of aluminum or glass reinforced plastics. Alloys commonly used for temperatures greater than 480 °C are the heat treatable types. Non-heat treatable metals such as austenitic stainless are used for lightly loaded parts for temperatures to 980 °C. Columbium, molybdenum, and tantalum are used at temperatures greater than 980 °C.

Titanium   Titanium honeycombs are competitive with stainless steels. Their use is somewhat limited by forming and machining difficulties.

Glass Reinforced Plastics   Many GRP formulations are available, most of which are made by impregnating glass cloth with resin and expanding it to honeycomb, curing, then coating with a resin. Cores are available in cell sizes from 5 to 10 mm and in thicknesses ranging from 1.5 to 45 mm. Density ranges from 50 to 160 kg/m³. Cores are available normally in slices precurved to specific contours or as blocks.

Paper Cores   Paper honeycomb is usually impregnated with a phenolic resin for strength, rigidity, and moisture resistance. Cell sizes vary between 8.5 and 30 mm, with core thicknesses from 6.4 to 150 mm. Depending on paper thickness and amount of impregnation, densities range from 24 to 64 kg/m³.

Bonding and Joining

Many types of adhesive are used in honeycomb panel manufacture: rubber based cements, combinations of thermosetting resins and elastomeric polymers, epoxy resins, epoxy – phenolic systems, and duplex types consisting of a supported film adhesive (made of a combination of thermosetting resin and an elastomeric polymer) on one side and a film of semi-liquid epoxy on the other. Adhesives must wet and adhere to both core and facing without causing deterioration or deformation. Shear, tensile, cleavage, and peel strengths available should be compared to job requirements. Adhesives vary in their capacity to withstand weather, moisture, chemicals, temperature extremes, impact loads, and fatigue. Resistance welding and brazing are common bonding processes with metal honeycombs.

Structural and environmental conditions will determine to a large extent which materials are used in honeycomb structures and their methods of joining. Generally, honeycomb structures cannot withstand compressive loads caused

146

by bolts, screws, and other mechanical fasteners. However, many special fasteners and inserts have been developed to overcome the problem. Inserts include wood, plastics, metal strips, and aluminum spacer inserts.

Creep, Impact, and Fatigue

Creep properties of honeycomb panels are primarily a function of the adhesive, service temperature, and time at temperature. With the proper selection of adhesive, most creep requirements are easily met. In some cases, concentrated impact loads—especially for floors—outweigh all other design considerations. Thicker facings are required to resist such loads. Panels can be designed with different facing thicknesses. Honeycomb structures are probably more resistant to fatigue than any other construction. In fatigue tests, panels usually fail at the attachment points.

Utilization

Applications are usually governed by advantages, disadvantages, and economics. Advantages include:

> high strength-to-weight ratio
> resistance to heat transfer and vibration
> use of nearly any structural material
> close tolerances
> high speed production
> ease of fabrication

Some disadvantages are:

> difficult to inspect and repair
> high cost of some honeycomb materials
> joining problems
> problems with adhesive bonds and shear strength

Commercial applications include wall panels, flooring for trailers, boat hulls, ship doors and bulkheads, table tops, pallets, truck panels, doors, stressed skin building, etc. Aircraft applications include wings, rotor blades, trailing and leading edges, doors, flooring, bulkheads, stabilizers, radomes, fuselage sections, elevators, and rudders. More recent applications have included gas turbine ducting, radiators, shock absorbers, electromagnet shielding, and noise suppressors.

Industrial Glasses

Glass first appeared in its natural form, known as obsidian, an opaque variety dating back to about 3000 BC. The Egyptians produced a translucent glass about 1500 BC which was used for ornaments. Modern glass is transparent and

147

has properties exploited in a variety of applications such as optical glass, safety glass, glass for directing illumination, glass tiles, and glass building blocks.

Large-scale production of glass makes use of regenerative furnaces similar to those used for making open hearth steel. The raw materials, sand, sodium carbonate, and lime, are charged at one end of the furnace. At the exit end there are devices for drawing the glass into sheet, tubing, and filaments or for converting the plastic mass into other shapes. Blowing is another method of shaping glass.

Following fabrication of the particular glass shape, it must be annealed or slowly cooled to prevent formation of residual stresses. If not slowly cooled initially, the glass is reheated to just below the softening temperature and then cooled slowly. Glass, particularly in sheet form, is sometimes strengthened by a heat treatment known as 'tempering'. The glass is heated to just below the softening temperature and is cooled in an air blast or oil bath. This places the outer portion in compression and the inner portion in tension, thus increasing resistance to bending failure. Glass, being only about 10% as strong in tension as compression fails in tension when bent. Hollow blocks for structural purposes are made in two parts which are welded together by heating and the use of special metal films.

Properties

Glass is amorphous material, said to be a super cooled liquid. Glass can be converted readily to its plastic state by heating to about 815 °C. Its distinguishing feature is transparency, hence its use in windows and optical devices. It is excellent in corrosion resistance, as it is not attacked by salts, acids (except hydrofluoric), alkalis, or organic substances. It is hard and wear resistant but lacks ductility. It has poor resistance to thermal shock and poor creep resistance at elevated temperature. It is different from metals in that it has good electrical resistivity. Its conductivity improves at high temperatures, hence it cannot be used as a high temperature resistor. Glass is a brittle material and at room temperature it is almost perfectly elastic.

The strength of glass depends on residual stresses which develop as a result of cooling or other thermal treatments, the surface condition, and surrounding atmosphere. A network of microscopic cracks covers the surface of a glass and these cracks are a major factor in reducing its strength. Water and water vapor are mildly corrosive when absorbed within such cracks and serve further to reduce strength. Where the area is small, such as in glass fiber, tensile strength can be extremely high, but for larger specimens is reduced appreciably because of flaws. Except for specially tailored glasses, visible light up to 90% is transmitted. Glass, however, is usually opaque to ultraviolet light and translucent to infrared light.

Types of Glass

Lime-soda Glasses These contain silica or silicon dioxide as the major

148

ingredient. Pure silica produces a high quality cast glass but its poor workability, high melting point, and its tendency to trap air bubbles limits its use. It has the best thermal and chemical shock resistance of all glasses. With increasing amounts of sodium oxide, the melting point and chemical resistance are lowered. Introduction of calcium oxide increases resistance to water, chemicals, and abrasion. Lime-soda glass is used in windows and doors.

Borosilicate Glasses These use boric anhydride instead of lime. They have a low coefficient of expansion, and hence good spall resistance. Pyrex is a trade name for this type of glass. It is used in cooking utensils and astronomical telescopes.

Leaded Glasses These contain as much as 92% lead oxide. They are used for γ-ray and X-ray protection occurring in electronic tubes. They are sometimes called flint glasses because of their decorative and optical uses resulting from their high index of refraction.

Phosphate Glass This is useful in the transmission of a wide range of wavelengths. It transmits about 80% of the entire ultraviolet spectrum.

Fused Silica or Quartz Glass This is made from commercially pure silica and has a very low coefficient of expansion.

Pyrocerams These are crystalline glasses which have the properties of ceramics.

Uses

Glass has uses in all fields of engineering. In civil engineering, it is a structural material and is used extensively in buildings. In mechanical engineering, it is used as glass pipe, glass fiber insulation, and glass springs. In electrical applications, glasses are used as insulators and tube envelopes where glass – metal bonds are relied upon to maintain vacuum.

Properties of thermosets can be greatly improved with the use of glass fillers. Fibrous glass improves impact strength without loss of electrical properties. They have good resistance to high temperature. Preformed sheets and planks with glass filling offer high tensile and flexural strength and excellent moldability. Recent developments include glass deep submergence vehicles, high modulus beryllium glasses, and low iron content glasses used in solar heating systems.

Natural and Synthetic Fibers

A fiber can be defined as a form of matter of small cross-sectional area and with length greatly exceeding width. It the length is so long that it may be regarded as infinite, it is referred to as a continuous filament. A short length of no more than about 100 mm is called a staple fiber. Man-made fibers are produced as continuous filaments which may then be cut up to make staple fiber. Wool and cotton are natural staple fibers while silk is a natural continuous filament.

Fibers, both natural and synthetic, are either organic polymers or inorganic substances which are crystalline or polymeric. Examples are cotton, wool, rayon, nylon, glass, or asbestos, and others as shown in Table 11. Fibers may be laid out as a mat and bonded by an adhesive or by heat setting, the resultant product being called a man-woven or bonded fabric. Yarns can be made by gathering and twisting several fibers into a single string-like element which has good axial strength from interfiber friction. Fibers are used widely as reinforcement in plastics to strengthen and stiffen them and to increase tear resistance.

Natural Fibers

Cellulose Cellulose is the reinforcing fiber found in all vegetation. In its crystalline form, it is the main structure of wood fibers and various natural textile fibers. The main drawbacks of such fibers are that they absorb moisture, which causes swelling, increases weight, decreases mechanical properties, and increases creep. They are also vulnerable to attack by insects and fungi. The chief vegetable fibers are of four types: the soft hairs which surround seeds, tough fibers called 'basts' which grow between bark and stems of trees and plants, vascular fibers as found in leaves, and the entire stems of grasses.

Only two seed hairs have commercial importance, cotton and kapok. Cotton is the most widely used vegetable fiber. It is strong, but not highly stretchable because of its short fibers. Partly because of low elongation, cotton fiber does not recover well after stretching and cotton fabrics do not exhibit good wrinkle resistance. Cotton has the advantage of low price and ready availability and can be processed efficiently to yarns and fabrics. Kapok is used as a stuffing filler. It is buoyant and impermeable to water and is used in life preservers and in upholstery stuffing. The coconut fiber, called coir, is used for bristles, door mats, and outdoor carpets. Brush fibers are obtained from the stems of various palms.

Some examples of bast fibers are flax (which is woven into linen), hemp, jute, and ramie. These are cheap fibers used in sacking, webbing, industrial cloths, backings for rugs and carpets, and for twine. Vascular fibers are used entirely for cordage and include manila, sisal, and yucca. Entire stems of some grasses and straws are used as fibers in weaving of hats, matting, and furniture pieces.

Animal Fibers There are two basic types of animal-based fibers. First, silk which is spun in continuous filaments from the abdomen of various species of insects, spiders, and the silk worm. Second, the hair, furs, and wools used as a protective covering on mammals.

All animal fibers are complex proteins and can be attacked and ultimately dissolved by the action of alkali, although they have fair resistance to acids. Hair and silk are resistant to other chemicals whereas wool is bleached. Wool has good resistance to heat to 100 °C, silk to 170 °C. All burn when exposed to flame. Wool and hair have average resistance to micro-organisms and silk has good resistance. Fibers of hair and wool are not continuous and must be spun into thread or yarn for woven textile fabrics or alternatively they may be felted. The chief hair fiber is sheep's wool. Wool structure differs from other hairs in that the fibers have a coat of overlapping scales, whereas hairs are smooth. This gives wool the property of a directionally dependent coefficient of friction which encourages undirectional fiber migration and entanglement when a fiber mass is worked in a wet environment. This is called felting.

Other fibers derived from the hair or furs of animals are angora and cashmere from the goat, mohair, mink, badger, and beaver fur. These are used in apparel. Cattle, goat, and horse hair are used for stuffing and wadding. Hog bristles are used in brushes. True silk is made only from the silkworm. It is almost as strong as nylon and has almost equal elongation and toughness. Wild silk, which is produced by several related species, is rectangular in section and irregular. The high cost of silk restricts its use to expensive fabrics because it has excellent properties of strength, drape, warmth, softness, and is smooth to handle.

Mineral Fibers   Asbestos is the only natural mineral fiber. Fibers are obtained from rock by repeated crushing and air suction removal, and are cleaned by screening. Asbestos fibers are hydrated silicates formed by high pressure hydro-thermal reaction. They have resistance to acids and excellent resistance to alkalis. They are not flammable and have excellent heat resistance to temperatures of 810 °C. Asbestos fibers may be woven or felted into cloth and sheets. Asbestos fabric is used in certain types of brake linings, gaskets and packing, blankets, and fire resistant clothing. There are health hazards associated with asbestos ingestion.

Regenerated Cellulose Fibers   These fibers include viscose (rayon), cellulose acetate, and fluorocarbons (Teflon). Rayon is an extremely versatile fiber and is widely used in textiles, especially when blended with synthetic fibers. Uses of rayon include all types of clothing, domestic curtains, upholstery, covers, blankets, sheets, carpets, nonwoven fabrics, disposables, paper, and medical and surgical applications. The high tenacity fibers have tire cord use (especially in radial tires) and industrial uses in hoses, tapes, and belts.

Cellulose acetate fibers are made by dry spinning a solution of the polymer in dichloromethane into hot air. They are thermoplastic, burn slowly, are resistant to micro-organisms and melt at about 260 °C. Cellulose acetate has fair resistance to acids and good resistance to alkalis. It has wide usage in dresswear, especially for drip dry clothing. Fluorocarbon fibers are inert to micro-organisms, acids, and alkalis, and have good resistance to chemicals. They are self-extinguishing and have a maximum service temperature of 290 °C.

Synthetic Fibers

Among others, synthetic polymer fibers include polyacrylonitrile, poly-amides, polyesters, phenolics, aramids, and inorganic types.

151

Table 11   Industrial Fibers Profile

Property	Cellulose Natural[1]	Cellulose Regenerated	Polyamide Regular tenacity	Polyamide High tenacity	Polyamide High Temperature[2]	Polyester	Acrylic[3]
Breaking tenacity (g/denier)	3 – 5	1 – 5	3 – 6	6 – 9	5.3	3 – 6	2 – 4
Tensile strength (MPa)	410 – 680	135 – 680	300 – 600	600 – 890	650	340 – 680	200 – 400
Breaking elongation (%)	3 – 10	10 – 30	25 – 40	15 – 25	22	15 – 25	25 – 45
Initial modulus (g/denier)	35 – 40	50 – 70	30 – 40	35 – 40	140	80 – 100	40 – 80
Initial modulus (GPa)	4.8 – 5.5	6.8 – 8.9	2.7 – 4.1	3.4 – 4.1	17.1	9.6 – 12.3	4.1 – 8.2
Specific gravity	1.5	1.5	1.1	1.1	1.4	1.4	1.2
Strain recovery (% recovery from % strain)	45/5	50/5	86/5	89/3	—	63/5	62/5
Creep	Low	Med	High	High	High	Med	Med
Energy to break	Low	Med – high	High	High	Med – high	Med – high	Med
Abrasion resistance	Med	Low – med	High	High	Med – high	Med – high	Med
Flex endurance	Med – high	Med – high	High	High	High	High	High
Maximum usable temperature (temperature at which 50% of fiber strength lost in air) (°C)	120	205 – 260	175	175	230	205	120
Flammability	High	High	Med	Med	Low	Med	High
Resistance to weather	Med	Med	Med	Med	Med	Med – high	High
Resistance to acids	Low	Low	Med	Med	Med – high	High	Med
Resistance to alkalis	Med	Med	High	High	High	Med	Med
Resistance to solvent	High	High	High	High	High	High	High
Relative price	Low	Low	Low	Low	High	Med	Low

[1] Dry, cotton (staple).
[2] Nomex.
[3] Staple.
[4] Commercial yarn.
[5] Teflon.
[6] E and S glass.
[7] Carbon and graphite yarn.
[8] Chrysolite ashestos.
[9] 0.013 mm diameter fibers.
[10] 0.13 mm diameter fiber.

Mod- acrylic[4]	Poly propylene	Fluoro carbon[5]	Glass[6]	Carbon aceous Residue[7]	Mineral[8]	Steel[9]	Boron[10]
2.5 – 3	3 – 15[4]	1.2 – 1.4	15 – 20	2.4 – 16	1.3 – 13	1.5 – 2.5	12 – 18
290 – 680	240 – 1200	220 – 260	3300 – 4400	270 – 2200	270 – 2700	1000 – 1700	2700 – 4100
30 – 40	20 – 35	15 – 35	3 – 5	0.5 – 2.0	1 – 3	1 – 7	0.7 – 1.0
40	30	12 – 14	310 – 380	360 – 2400	130 – 780	290	1650 – 1800
4.8	2.4	2.0 – 2.7	68 – 82	40 – 340[11]	28 – 170	205	375 – 410
1.3	0.9	2.1	2.5	1.3 – 1.6	2.25 – 2.75	7.9 – 8.4	2.6
80 – 97/2	90 – 100/5	—	100/3	100	100	100/0.5	100
Med	High	Low	Zero	Zero[12]	Zero	Low	Zero
Med	Med – high	Low – med	Low	Low[12]	Low	Low	Low
Low – med	Med – high	High	Poor	Poor[12]	Low	Med	—
Med – high	High	High	Low	Low[12]	Low	Med	—
—	60	80	420 – 450	480	590	535 – 760	535
Low	Med	Low	Non-flammable	Low	Non-flammable	Non-flammable	Non-flammable
Med – high	Med	High	High	High	High	High	High
High	High	High	Med	High	Low	Med	High
High	High	High	Med	High	Med	High	High
Low	High	High	High	High	High	High	
Low	Low – med	High	Low – high	High	Low	High	High

[11] 680 GPa has been achieved in laboratory.

[12] Noncoated fibers.

Reproduced by courtesy of Fabric Research Laboratories, Dedham, Mass., USA (1968).

Polyacryonitriles Staple acrylic fibers, being soft and resilient, are used as a substitute for wool. Fabrics made from these, trade named 'Orlon', show good crease resistance and crease retention (permanent press pleats). Rot and light resistance properties make acrylic fibers suitable for outdoor applications and for fine and heavy duty fabrics.

Polyesters Dacron and Terylene are light- and moderate-weight polyester fibers. They are quick drying and make crease resistant fabrics. They are used as fatigue resistant reinforcements for tires, Vee-belts, and conveyor belts. They are used widely in fishing nets, tarpaulins, and in sail cloths.

Polyamides The most well known of this group is nylon. Nylons are characterized by good strength, toughness, and elasticity. They have the appearance and luster of silk and tensile strength greater than that of wool, silk, rayon, or cotton. Nylons have nearly replaced silk in the manufacture of hosiery, sleepwear, underwear, and raincoats. Nylon fabrics are water resistant, dry quickly, and do not require ironing. Parachutes, insect screens, and fishing nets and lines also use nylon fibers. They have outstanding abrasion resistance—better than any other fiber—and are used in carpets alone or as blending materials. 'Nomex' is a recently developed nylon structural fiber.

Inorganics Molten glass can be extruded to form a very versatile fiber. It has very high strength and a reasonably high modulus. These properties and low moisture absorption give rise to fabrics which have excellent dimensional stability. Glass is not degraded by exposure to sunlight, has very high resistance to chemicals, and will not burn.

Glass fiber is used widely for insulation, for filtration of hot gases and reactive chemical solutions, and for plastic reinforcement. Its poor abrasion resistance, low extensibility, and problems of dyeing glass fabric seriously restrict its use in apparel fabrics. Other inorganic fibers include a variety of metals.

Carbon fibers are a fairly new inorganic fiber development. They are primarily used in composite materials where unique properties of high in-service temperature, low thermal expansion, strength, and very high stiffness are needed. They are used in aircraft and spacecraft structures, for compressor blades in jet engines, rotor blades, bearings, pressure vessels, and high speed weaving machinery parts. Electric power cables of aluminum and carbon fibers have been studied as a means of providing high conductivity and high strength.

References

CHAPMAN, C.B. *Fibers*, Butterworth, London, 1974
PARRATT, N.J. *Fiber Reinforced Materials Technology*, Van Nostrand Reinhold, New York, 1972

Papers

As a medium of writing, paper has played an important part in the development of civilization and culture. However, writing paper today accounts for only a small proportion of the paper used. Paper is a material made up of many small fibers bonded together, most commonly cellulosic in nature. Paper may include sheet materials produced from other types of fibers such as mineral or synthetic. A major attribute of paper is its versatility. End properties can be closely controlled by selection of the fiber type and size, pulp processing method, web forming operation, and treatments which can be applied subsequently.

The crude fibrous wood pulps from which the papers are made are of two types: mechanical wood pulps and chemical wood pulps. The mechanical types include ground wood, defibrated wood pulp, and exploded wood pulp. Chemical wood pulps are produced by 'cooking' the fibrous material in various chemicals to provide desired physical characteristics and include sulfide pulps, neutral or monosulfite pulps, sulfate as kraft pulps, soda pulps, semichemical pulps, and screenings.

Applications

Most papers are used in the packaging and building industries. Packaging paper is either used alone or combined with plastics, resinous materials, or aluminum foil. Packaging paper is divided into three groups:

1. consumer-type packages such as small bags, where decorative and other functional properties such as moisture and grease resistance are more important than strength;
2. bulk shipping containers and other heavy duty applications where strength is the most important consideration—decorative purposes are not so important, although a printing surface is often required;
3. other types.

Group 1

Grease-proof paper   A dense, highly hydrated, uncalendered paper made from pulp produced by the sulfite process. It has good grease- and odor-proof qualities and is used extensively in food packaging such as carton liners and candy wrappers.

Glassine Paper   Made from highly calendered grease-proof paper. It is very dense, smooth, and transparent or semi-transparent. Glassine paper has better grease- and odor-proof qualities than grease-proof paper. Surface coating or laminates can be applied and it is used for envelope windows and sanitary wrappings.

Vegetable Parchment Paper   'Parchmentized' paper obtained from a carefully controlled process using sulfuric acid. Depending on the pulp, manufacturing process, and acid treatment, the paper can have varying properties but can be grease-proof and have a high wet strength. It is used for wrapping butter,

155

margarine, meats, and moist food products. It is also a liner for lard and butter tubs.

Bag or wrapping papers   Made by either the sulfate or sulfite process and used as a base stock for coatings and plastic films. Coatings used on these include wax or hot melt coatings with a wax base, wax laminated paper, lacquer or solvent coatings, laminated fibers, extruded polyethylene coatings, and aluminum.

Group 2

Kraft paper is used mostly in the packaging field because of its strength and low cost. Made from pulp produced by the sulfate process, it is available in a range of colors and finishes, the most common being brown or 'natural kraft'. It may contain urea or melamine formaldehyde to give wet strength properties. Group 2 papers include the following:

Treated Kraft   This is modified to give moisture-proof, grease-proof, and alkali resistant properties. These can be reinforced to increase strength by coatings or laminations.

Asphalt Laminated Kraft   In this case two sheets are laminated with asphalt to give moderately effective moisture proofness at low cost. They are used as multiwall bags and wrapping paper, but are vulnerable to creasing and become stiff at low temperatures.

Polyethylene Kraft   Here an extruded polyethylene coating is applied to give an excellent moisture barrier, acid and alkali resistance, prevention of fiber contamination, and grease-proofness.

Waxed Kraft   These are classed as 'dry' or 'wet'. 'Dry waxed' has a uniform impregnation of wax or wax – oil mixture, not a wax coating. It is an improvement over untreated paper for moisture and grease protection. 'Wet waxed' has a coating of wax and is susceptible to creasing. It is used frequently for the inner plys of multiwall bags for frozen vegetables.

Reinforced Kraft   These types are made similarly to asphalt kraft, but have a greater amount of asphalt laminate. They may be reinforced with materials such as glass fiber embedded in the asphalt interlayer. Most reinforced papers are custom made.

Group 3

Coatings are required for many types of paper to enable printing of designs. Quality white clay (koalin) in a starch binder is the usual coating. Also used are titanium dioxide and porcelain enamel for whiteness. Building papers and roofing felts are made from paper board coated with asphalt. Types include the following:

156

Printing paper   The paper used for books, newsprint, and magazines is the next largest use after packaging. Writing paper can be made from rags, both cotton and linen. The best paper is 50 – 100% rag paper. Specialty papers include those for legal documents, insurance policies, share certificates, bonds and bank notes, drafting and blueprint paper, high grade letterhead bond paper and special stationary, and cigarette, carbon, and typewriter paper.

Electrical papers   These are used for electrical insulation. Insulation paper is usually kraft paper coated on both sides with black or yellow varnish.

Capacitor paper   This is used as a dielectric in capacitors and is made from Swedish spruce sulfate pulp. It is highly purified, nearly transparent, and extremely thin, but is strong and tough.

Absorbent papers   These include blotting and filter papers made from spongy fibers or loosely felted paper. Most oil, fuel, and air filters are made with it.

Building papers   These can be used as structural sandwich material where papers are impregnated with phenolics and formed into honeycomb cores. Building paper used for house sheathing is a heavy kraft paper.

Plywoods

Plywood is the product obtained when thin layers of timber, i.e., veneers, are glued together with the grain of one ply running at right angles to the grain of the adjacent ply. The plys or layers are usually arranged in an odd number, so as to preserve a balanced structure and minimize warp due to load, temperature, or humidity changes. There are three-ply, five-ply or a greater number of multi-ply constructions available.

There are three major techniques in veneer production: rotary peeling, slicing, and semi-rotary cutting. Rotary veneer is processed on a 'veneer lathe'. In this machine the log is turned on a knife-nosebar assembly which is fed towards the centre of the log at a constant rate, resulting in a continuous ribbon of veneer of uniform thickness. Rotary veneer manufacture represents about 95% of the total produced. Sliced veneer is processed on a 'slicer'. Veneers manufactured by slicing are used for decorative purposes because of the figure or pattern obtained. Semi-rotary veneer is produced on a rotary veneer lathe. It is a combination of slicing and rotary peeling and is used only to obtain decorative veneer.

It is generally recognized that the adhesive or bonding agent used in the manufacture of plywood is of the utmost importance. Its characteristics determine the nature of the final product. Adhesives used currently are based on synthetic resins and are all themosets with some limited use of crosslinked PVA elastomers. The principal difference between the adhesives is the degree to which they are water-proof. A series of bond tests ranging from Type A to Type D, in descending order of permanency, is defined by standards. Plywood manufactured to Type A bond has a glueline which will not deteriorate under the

157

action of water, extremes of hot or cold, or prolonged stress. It is readily recognized by the black or deep red color of the glueline. Marine, exterior, and structural plywoods have Type A bond. Type B bond plywood is incorporated within the exterior standard. However, this adhesive will in time break down under long-term stress. The glueline, therefore, cannot be termed fully permanent. For example, the standard accepts that Type B bond plywood can be used for concrete formwork, but with limited life expectancy. Plywoods manufactured to Types C and D bonds are for interior use only. These gluelines can easily be recognized by their light color. Urea resin glues are used.

Properties

Traditionally, the decorative properties of plywoods have been used in furniture and feature wall panelling. The simple technique of bonding veneers is still the most economical and effective means of reproducing the beauty of figured timber in flat, stable panels. Since timber is about 25 – 45 times stronger along the grain than across the grain, plywood with the grain of alternate laminations at right angles tends to equalize the strength in all directions. Due to its construction, the section shear strength at right angles to the plane of the veneers is twice that of the parent timber of the same thickness. This structural characteristic is exploited in plywood box beams where the webs carry high shear forces near supports. Plywood has exceptionally high strength-to-weight and stiffness-to-weight ratios. Dimensional stability is high for moisture content and temperature changes. Plywoods offer the same insulation properties as the wood from which they are obtained. The combination of strength, stability, and good insulation properties make plywood a versatile building material. In addition to retaining some of the desirable features of solid wood, plywood improves on some of the less desirable properties such as expansion and shrinkage across the grain. Resilience, fatigue, and impact resistance are also inherent in plywood.

When plywood is subject to concentrated loads the two-way laminated construction assists in distributing the pressure over a large area, thus giving it excellent resistance to puncture. Because of the two-way construction, there is no potential cleavage line when pierced by sharp objects. Damage is normally limited to local tearing of the face and splitting does not occur. For this reason nails can be placed close to the edges without fear of splitting. However, it is not considered good practice to nail into end grains, as plywood is weak in that direction. The glue bond will not part parallel to the veneers, but wood fibers can split easily by shearing action along end grain veneers.

Products and End Uses

It has been demonstrated that there are numerous applications of rugged low-cost structural plywoods. They not only out-perform other wood based panel products in mechanical properties, but also in consumption *per capita* and in the variety of end uses. Some notable construction uses for structural plywoods are subfloors, concrete formwork, sheds, silos, trailer decks and sides, industrial storage shelving, targets and ammunition boxes, and refrigerated cars.

Marine plywood is designed to cater for two distinct marine purposes. First (for high strength, high durability), thick marine plywoods are specified for ocean-going yachts, naval craft, police launches, barges, and commercial trawlers. The second common marine application is where medium strength but extreme lightness is required, such as for racing class boats. These generally employ thicknesses from 3 to 10 mm and are popular with home boat builders.

The strength and light weight of plywood, its availability in large sheets with a range of types and grades, its resistance to warping and splitting, its ease of working and bending and being formed result in a variety of uses. Marine plywood can be specified for nonmarine structural applications requiring thin, lightweight material of particularly high strength where sanded decorative faces are sometimes required. Box beams, light aircraft, water cooling towers, and structural models are applications which require thin plywoods of high shear strength. It is often the case that marine plywood will be specified in the belief that it is the strongest available. In many cases structural plywoods are adequate and of lower cost.

Exterior plywoods can be distinguished usually by their thin face veneers which have a highly paintable grade or resin film overlay on one side. They can incorporate either a Type A or a Type B bond. They are used for garage doors, wall cladding, roof tiles, blackboards, garden furniture, and dog kennels. They are also find interior uses where moisture prevails such as in bathrooms, laundries, shelving, and sink bases.

Standard construction plywoods can be overlaid with a variety of materials to give improved weathering resistance or abrasion resistance. Many types of over-lays are bonded to plywood for special applications. Phenolic resins, metals, polyester resins, glass fiber, and bituminous water-proof membranes have been used successfully. Aluminum, lead, copper, and stainless or galvanized steel can be pressed on to plywoods with the use of adhesives. Such panels are used in hospitals, counter tops, and wall cladding where surfaces must be decorative, hygenic, and easy to clean.

In the above applications, the ability to order large panels in exact sizes saves much material waste and enables the user to calculate accurately the quantities required.

References

Plywood Manual, Plywood Association of Australia Ltd, 1976

HANLEY, D.P. McDOWALL, C.G. AND LYNGCOLN, K.J. Optimization of Plywood Sheeting for Transverse Loads, *Australian Forest Industries Journal*, May 1976

HANLEY, D.P. Optimization of Plywood Sheeting for Buckling Loads, *Australian Forest Industries Journal*, August 1976

HANLEY, D.P. *CIAE Study Leave Report*, A review of world developments in the plywood industry, January 1978

AS 085: Pinus Structural Plywood
AS 086: Plywood for Marine Craft
AS 087: Plywood for Exterior Use
AS 088: Plywood for Interior Use
AS 090: Methods of Testing Plywood

Structural Glulam

The term 'glulam' stands for 'glue laminated timber' and refers to a form of construction in which a number of laminations are arranged parallel to the axis of a member, the individual boards comprising the laminations being assembled and glued together to form a member which functions as a single unit. Thus the 'grain' of glulam is parallel, in constrast to plywood in which the grain alternates with each lamina. 'Microlam' is the trade name of a recent product composed of thin veneers; 'Presslam' is that for a thick laminate product.

To join two or more pieces of timber by gluing, the following procedure is necessary:

1. Timber be stress-graded in accordance with design requirements.
2. It must be kiln dried to a moisture content of normally $12 \pm 3\%$.
3. It should be machined on the gluing surface within 48 h prior to gluing.
4. The adhesive should be applied over the entire surface, preferably by mechanical means and the surfaces brought into close contact within the specified assembly period.
5. Clamping pressure is applied during the closed assembly period.
6. Temperature and pressure are in accordance with relevant standards and/or adhesive manufacturer's recommendations.
7. A suitable post-curing period is allowed after clamping.

Major attention is given to gluing and joining as next described.

Gluing

Three main types of glue are used for glulam construction. These are phenolic resins, casein glues, and urea adhesives. Phenolic resins possess qualities desirable in a structural adhesive, including water resistance, but they are expensive and difficult to apply and require stringent temperature controls and timber conditioning. Casein glues are easier to apply, require lower temperatures, and are not as sensitive to timber conditioning. They are limited to internal uses, although they have some degree of water repellance. These glues are less costly than phenolics and are not used widely in structural applications. Urea adhesives have been used extensively since glulam was first introduced. Unfortunately they break down when subjected to temperatures greater than 38 °C for long periods and are not water resistant. Melamines and resorcinols are other adhesives used, often in combination with phenolics to reduce cost.

Joining

There are three joint types most used in glulam: scraf, butt, and finger joints. There are certain difficulties with the use of scarf joints, both in manufacture and assembly. If the laminates are not precisely located, and overlapping or insufficient lapping occurs at the splice, low strengths result. It may, however, be cheaper to use butt joints and provide more timber to compensate for the weakening effect than to use properly fabricated but more costly scarf joints. Scarf joints are essential in sharply curved members and/or when appearance is important or weight must be kept to a minimum.

Butt joints are less costly and are also less efficient than scarf joints. They cannot transfer stress from one lamina to another, either in tension or compression. If butt joints are used, they should by systematically staggered so that there is never more than one on any crosssection.

Finger joints are widely used to produce long lengths of timber from short pieces, thus reducing costs and conserving resources. There are several factors which affect the strength of finger joints, including slope of the fingers, width of the finger tips, length of the fingers, and the pressure used to force the joint ends together. High speed finger jointing machines provide controls for these variables.

Advantages and Limitations

One of the great advantages of laminated timber is that it may be fabricated to straight and curved shapes and to unusually large cross sections and lengths. The use of seasoned timber and the dispersion of defects results in higher allowable stresses than can be obtained in solid sawn timber. Some of the advantages of glulam are as follows:

1. Structural simplicity has appeal to the designer. Lamination allows timber to be formed to many dimensions and shapes. Resulting forms are clean, efficient, and often beautiful. If required, as is often the case in church interiors, a fine finish can be given to the structural member.
2. Small sizes of lower grades of timber can be utilized in the construction of large members.
3. Timber of higher stress grades may be placed in those parts of the built-up member where stresses are greatest, the low-grade material being used where the stresses are lowest.
4. A laminated member of large cross-section is more fire resistant than construction with many small exposed individual pieces.
5. Curved shapes with varying cross-sections may be constructed.
6. Laminated beams may be cambered during construction to eliminate sag that frequently accompanies the seasoning and loading of a straight monolithic beam.
7. Proper gluing enables the laminates to act as a unit more effectively than is possible with nails, bolts, or other mechanical fasteners.
8. Laminated construction simplifies the design of fixed joints in members such as portal frames, thereby gaining economy.

161

9. Glulam has a special appeal to builders because the seasoned laminations result in less shrinkage to spoil plaster and other joined materials.
10. The resistance of timber to corrosion by steam or acrid fumes offers another advantage in factory construction.

Although laminated beams offer certain advantages, the cost of preparation usually exceeds that of a solid green member. If green timber is acceptable, it can probably be obtained in less time than that required for glulam. Laminated members must be fabricated in a plant and the member sizes may be limited by transportation facilities. The laminating process requires special equipment and fabricating skills that are not required for the production of solid green timbers. Since glulam is not usually mass produced, there can be delays in construction.

References

ANDREWS, H.J. *An Introduction to Timber Engineering*, 1st edn, Pergamon Press, London, 1967
CHUGG, W.A. *Glulam*, Ernest Benn, London, 1964
Parker, H. *Simplified Design of Structural Timber*, 2nd edn, John Wiley, London, 1963
PEARSON, R., KLOOT, H. and BOYD, J., *Timber Engineering Design Hand-book*, 2nd edn, Jacaranda Press, Melbourne, 1962
WALLIS, N.K., *Australian Timber Handbook*, 3rd edn, Angus and Robertson, Sydney, 1970
Wood World, Volume 9, Number 2, ATJ Promotions for the Timber Development Association, Sydney
Australian Standards Association, AS 1328 – 1972; Glued Laminated Structural Timber, 5 July 1972

Wood and Wood Base Composites

Man's association with wood dates back to primitive times. Wood is a very complex material because it is the product of the metabolism of a living organism, a tree, and because its properties are subject to wide variation brought about by factors affecting tree growth. Furthermore, wood is a product of not one, but of many tree species and thus each kind of wood exhibits its own anatomical, physical, and chemical characteristics.

Properties Common to All Wood

All woods possess certain characteristics in common:

1. All tree stems have a predominantly vertical arrangement and radial symmetry.
2. Wood is cellulosic in structure and chemical composition of wood cells comprises cellulose, noncellulosic carbohydrates, and lignin.

3. All wood is anisotropic in nature, i.e., it exhibits different physical properties when tested along three major axes.
4. Wood is a hygroscopic substance, i.e., it loses and gains moisture as a result of changes in atmospheric humidity and temperature. Because of its anisotropic nature, moisture variations produce dimensional changes which are unequal in three directions.
5. Wood is susceptible to attack by fungi, micro-organisms, and insects. It is also flammable, especially when dry.

Characteristics of Certain Species

Woods of different species may vary widely in appearance and in physical and chemical properties. Obvious variations are in color, texture and figure, density, hardness, strength properties, degree of dimensional changes, insulating properties, machining, gluing, finishing characteristics, ability to hold fasteners, durability, and reaction to chemicals. These differences may be traced to anatomical structure. Almost infinite variations and combinations of properties of woods present unusual opportunity for matching materials with particular use requirements.

Extent of Variability Within a Species

Variability exists in trees of the same species. This variability may occur from tree to tree, may be evident in the wood taken from various places in the same tree, or may even be found within a given piece of wood. Such diversity in properties is largely a result of the growth pattern of trees as well as environmental influences such as climate, soil, moisture and growing space. Genetic factors also play an important role.

Wood as an Industrial Raw Material

The uses of wood are many and may be broadly classified as:

  fuel
  pulp and paper (including derived products such as cellulose filaments and
      films, plastics, and explosives)
  chemicals
  structural applications

Wood is an important source of energy. In technologically less advanced countries it may still be the exclusive source of domestic and industrial heat. As a result, it is becoming scarce in some developing countries. There is no other natural substance than can meet the ever increasing demands for paper and other pulp products. At present it seems unlikely that a synthetic material could be produced economically to rival wood as a source of pulp. Many chemicals can be obtained from wood, but only a few are commercially important at the present time. Wood finds its most extensive use in construction, in furniture, and in transportation containers.

Wood as a Construction Material

The characteristics that account for the versatility of wood are many. It may be cut and worked into various shapes with hand tools and machinery. It lends itself well to conversion in a plant and on-site fabrication. Conventional wood-frame construction remains competitive with any method of housing construction. Wood can be joined with nails, screws, bolts, or connectors, using simple tools, and produce strong joints. It can be fastened with adhesives which produce continuous bonds enabling full shear strength to be developed.

Although wood undergoes dimensional changes with variations in moisture content, such changes are relatively insignificant in the grain direction, that direction being the most important in structures. Dimensional changes that take place as a result of temperature are less significant in wood construction than in metal structural members. When heated, wood expands across the grain as much as metals, but only little in the longitudinal direction. Moreover, the increase in dimensions with temperature is balanced frequently by shrinkage caused by drying with a corresponding increase in strength. There is no such compensating effect in metal structural members which may expand and lose strength progressively when heated.

Wood is a combustible material but loses its strength only gradually under fire conditions if used in large enough sizes. When subjected to extreme heat, wood decomposes with the evolution of combustible gases and tarry substances, leaving behind a charcoal residue. This process is known as a pyrolysis reaction and retards the decomposition to the interior of the beam. Charcoal, which forms on the surface, is an excellent heat insulating material until it reaches the glowing stage. Because of the self-insulating property, the strength of a burning beam is reduced gradually, in proportion to the section converted to charcoal.

Many woods are surprisingly durable. Decay and insect damage can be largely eliminated by sound methods of design and by using properly seasoned wood. Where biological wood-destroying agencies are difficult to control, wood can be treated by impregnation with preservatives. Wood does not corrode. Its major constituents are quite inert to the action of many chemicals. It is therefore used in many applications where resistance to the disintegrating action of chemicals and to corrosion are important. When unprotected wood is exposed to atmospheric conditions it will weather. Most exterior applications, however, make use of protective coatings such as paints.

Because of its fibrous structure and the presence of entrapped air, wood has excellent insulating properties. The thermal conductivity through common brick is six times that of wood and eight times as great through a glass window. Wood insulation is effective throughout the year, against cold and heat.

Wood when reasonably dry is a poor conductor of electricity. This is important in uses such as powerline poles and may be significant in dwellings by minimizing the danger from broken or exposed wires coming into contact with structural members. Because of the nature of the cell wall and its distribution as

164

a system of thin walled tubes, wood possesses excellent flexural rigidity and outstanding properties in bonding.

Wood is an excellent energy absorption medium, which makes it an excellent material for floors and applications where energy absorption is important. Wood also possesses excellent vibration damping characteristics which is important in structures subjected to dynamic loads. Wood also possesses appeal because of its surface pattern and infinite variation of grain, texture, and color. As well as its use in the normal state, wood is used in structural applications in the form of wood base compositions which include laminated forms and plywoods, wood – metal laminates, and fiber and particle base composites.

Laminated Forms

Wood laminating is a process of building up assemblies ('glulam') by laying and fastening two or more pieces of wood together so that the grain of each layer is parallel to the grain of all others. This is in contrast to plywood, wherein each layer is positioned so that its grain is perpendicular to that of the preceding ply. The laminae are generally fastened by adhesives or glues which are selected according to service requirements. Applications of laminated forms and plywoods include arches, beams, trusses, spring boards for swimming pools, boat frames, sporting equipment, and large industrial and domestic uses.

Wood – Metal Laminates

A wood or wood derivative core is put between metal facing sheets and adhesively bonded to form wood – metal laminates. The resulting structure is light in weight, has good strength, and high flexural rigidity. Wood – metal laminates are used extensively in architectural, building, and transportation fields for wall panels, partition panels, truck and trailer bodies, and shipping containers.

Hardboard

The fiber and particle base woods are of two types, those made from hardwoods and those from softwoods. These terms have no relation to the hardness or softness of the wood; for example, balsa is a hardwood. The types are based on cell structure.

Hardboard in Australia is almost exclusively made from Eucalyptus species which are cut and delivered to the mill in billets. The billets are fed into chipping machines where they are reduced to chips. The chips are ground in steam heated vessels called defibrators which turn them into pulp. Further refining is performed prior to sheet formation. Water-proofing agents are added to the pulp, which is then formed into a continuous wet mat by pumping over a wire screen. Initially the water drains from the screen by gravity, but later suction boxes are used to remove more water. The mat still on the screen is then passed through a roll press to further reduce water content.

The mat is then cut to the desired lengths and loaded into a large press where it is dried at temperature. Each mat is conveyed into the press on the top of a woven wire screen which imparts the 'screen back' to the board, a characteristic of hardboard made by the wet process. A function of the mat is to provide an escape path for water in the form of steam. No artificial binders are used in the manufacture of hardboard. Upon leaving the press, sheets are baked for some hours in a heat toughening chamber. Moisture is then added in humidifying chambers to minimize expansion and shrinkage from atmospheric conditions. The boards are then cut to size.

Hardboard comes in various forms for specific applications. Tempered hardboard is treated with oils to give added strength and resistance to moisture absorption. Molded hardboards have their faces embossed. Pegboard is a perforated hardboard and plastic faced hardboards are surfaced with a hard melamine surface to provide a range of colors and patterns.

Particleboard

Particleboard (or chipboard as it is known in the UK) is the wood composite made from softwoods. It is produced by bonding together wood particles with a synthetic resin adhesive under simultaneous application of heat and pressure. The raw materials for particleboard are forest thinnings removed during tree management. Thinnings are debarked and chemically treated to prevent mold growth. The debarked logs are cut into short lengths which are fed into flaking machines. The flakes are made into two sizes, a coarse flake for the core and a finer flake for the surfaces. The flakes are milled to reduce size and to retain uniformity. They are then dried to about 5% moisture content, weighed, and sprayed with a fine mist of urea formaldehyde glue. A mattress is formed with a layer of finer flakes outside the coarser ones forming the core. These are then sawn and weighed. A number of these mattresses are pressed simultaneoulsy between hot steel plates. After pressing, the board is allowed to cure for about a week and each side is then sanded to provide a uniform thickness.

Advantages of wood particleboards include:

availability in large sizes and various thickness
little materials wastage
dimensional stability
facing with veneers or laminated plastics
can be worked with hand and power tools

Disadvantages of particleboards include swelling, moisture absorption, easily damaged in handling, poor impact strength, and properties generally inferior to structural plywoods.

References

PANSKIN, A.J. and DE ZEEUW, C. *The Textbook of Wood Technology*, Volume 1, 3rd edn, McGraw-Hill, New York, 1970
DESCH, H.E. *Timber: Its Structure and Properties*, Macmillan, London, 1968

Australian Standards

AS 1684: Light Timber Framing Code
AS 1706: North-Eastern Australian Eucalypt Hardwoods
AS 1720: Timber Engineering Code
AS 1728: Types of Timber Surfaces

Wool and Synthetic Felts

Felts are one of the oldest engineering materials. With technological advances, their properties have been improved by supplementary treatments. Thus, mechanical uses of felts have expanded and they are important in mechanical engineering applications. By definition, felt is a fabric built up by the interlocking of fibers by a suitable combination of mechanical work, chemical action, moisture and heat, without spinning, weaving, or knitting. Types of fibers include wool, reprocessed and reused wool, and synthetics. These may be made with or without admixtures of animal, vegetable, and synthetic fibers. Felts typically are durable, resilient, do not ravel when cut, and may be soft or hard. Wool felt and synthetic felt have similar applications. but synthetics may be tailored for specific properties.

Wool Felts

The manufacture of wool felt entails fiber blending and picking, carding, hardening, fulling, scouring and neutralizing, dyeing, drying, shearing, and press finishing. Cross-carding of fibers is used to achieve strength in both length and breadth directions. The density of the felt and its characteristics may be controlled by fiber blending, and by the degree of felting in fulling and hardening. Processes used in fabrication include die cutting, stripping, coating, impregnating, laminating, cementing, machining, molding, shaping, drilling, and grinding.

Felt is available in sheet or roll form. Mechanical sheet felts* are fabricated to thicknesses from about 6 to 75 mm and with specific gravities from 0.25 to 0.68. Mechanical roll felts+ are fabricated to thicknesses from about 2 to 25 mm, lengths from 5 to 10 m, widths from 1.5 to 2 m, and with densities in the same range as sheet forms.

Properties Felts are described according to their texture and color and, in the case of wool felts, by their wool content (from 50 to 100%). Wool felts are available in white, black, and gray. Texture ranges from fine to medium-coarse. Splitting resistance increases with density and is independent of thickness. Under tension, felt develops a progressive reaction and resists increased loads. It is well suited to shaping by stretching or molding. Under compression, wool felt

---

* SAE Felt Standards adopted by ASTM under D-1144.

+ SAE Felt Standards adopted by ASTM under D-944.

does not crumble or flow and has low compression set. Wool fibers are natural springs and vibration isolation is an important function of felt. Wool felts have very low thermal conductivity and are used as thermal insulation. Sound absorption is high with wool – kapok felt. Higher density felts have greater sound absorption characteristics than lower density felts.

Wear resistance varies with density and fiber quality and load. High density felt is exceptionally wear resistant. The coefficient of friction of felt depends on its texture. Wicking rate of wool felt depends on fiber quality and felt construction, density, and the viscosity of the immersion fluid. Wick-rise rate and wicking distance increase with density and fluid viscosity and are independent of wick size.

Felts are efficient and economical for filtration of air, gases, and liquids. Felt filters have high permeability, long life, and low plugging rates. Extended immersion at 100 °C in distilled, tap, chlorinated, or sea water only slightly affects properties and redrying results in recovery of strength and splitting resistance. Wool felts may be considered ageless and resist moderate to strong chemical attack.

Applications   Vibration isolation, cushioning, and padding are natural uses of wool felt. Felt mountings are employed to dampen vibrations of floors and buildings which have precision instruments. They are used as an isolation medium for agitators, suction pumps, and centrifuges. Felt packing and pad mountings are used to protect machine parts and glassware from vibration during transportation. Cushioning for mechanical processes such as leather embossing is accomplished by using wool felts. Felts are also used as backing for the abrasive surface in sanding and grinding.

The coefficient of friction of felt, as well as its high abrasion resistance, enable it to be used as a base surfacing material for movable equipment and appliances such as telephones, electric fans, and chess pieces. Because of its capillary action, it may be employed in wick form to lubricate journal bearings, spindles, and other machines parts. Absorption, storage, and distribution of water, oil, and ink is accomplished by felt pads in printing equipment.

Felt is used as a seal to protect ball bearings from ingress of grit and moisture and to prevent lubricant leakage. The sealing effect is also exploited in felt pistons for grease guns, fountain pen, glass pipe gaskets, weather stripping, and drive shaft housing seals. Filters are used extensively as dust, fume, and pollen filters, as well as in filtration of petroleum, chemical, and food products. They are used in respirators and in filtering electroplating solutions, solvents, oils, paints, and lacquers. Felt is also used for sound-proofing and as thermal insulation.

Synthetic Felts

The interlocking of synthetic felts is attained by mechanical action, although in some cases heat and chemical action are utilized. Fibers used are cellulose, cellulose acetate, cellulose triacetate, polyester, polyamide, acrylic, and TFE fluorocarbon. Thus synthetic felts may be tailored for specific uses and have found more varied application than wool felts.

Properties   Superior properties may be obtained from synthetic felts including lower densities than wool felts, an infinite color range, quicker drying, improved chemical resistance, improved abrasive wear resistance, and lower moisture absorption.

Applications   Synthetic felts perform generally the same function as wool felts. Some specialty applications include the following:

1. *'Dynel'* felts, which have excellent resistance to acid attack and are used for battery separators and gaskets in corrosive environments.
2. *'Orlon'* felts, which also resist acids and microbiological attack and are used for filtration of dust and solvents.
3. *Nylon felts*, which are durable and chemical resistant and are used as filters, gaskets, and pads.
4. *Rayon viscose felts*, which are used for vibration isolation, sound absorption, and sealing in products such as sound-proofing, fuel and oil filter cartridges, water filters, and air conditioner seals.
5. *Dacron polyester felts* have an unusually high service temperature range (up to 170 °C), high sound absorption, chemical resistance, and air permeability, and are used in conveyor belting, thermal insulation, seals exposed to environmental attack, high temperature seals, wick lubrication, and air intake filters.
6. *Teflon fluorocarbon felts* have an exceptionally high temperature service range and chemical resistance, making them suitable for corrosion resistant gaskets.

Vibration isolation performance and flame resistance of synthetic felts are usually poorer than that of wool felts.

10
Protective Coatings and Nonstructural Materials

Coatings

Protective coatings are metallic, organic, or inorganic materials which are used for corrosion protection. They impart to a protected surface properties such as hardness and wear resistance, electrical and thermal conductivity, and oxidation resistance.

Protective coatings function by virtue of the interposition of a continuous physical barrier between the protected surface and its environment. Such barriers must be inert chemically to the environment under particular conditions of temperature and pressure and prevent penetration of the environment into the base material. Effectiveness of such barriers depends largely on their thickness, which may vary widely according to type of material, environment, and the required degree of protection.

Surface Preparation

Before a coating can be applied, the base metal surface must be thoroughly cleaned. Maximum coating adhesion results only if the surface is free of impurities and extraneous matter. The structure of the base metal surface also affects adhesion of the coating. Methods of surface cleaning depend on the type of impurities present and the base material. Some methods are as follows:

1. *Solvent cleaning* is used to remove oil and grease from articles prior to coating.
2. *Alkali cleaning* is used to clean the surface and is particularly suitable for removing old paints. The article to be coated must be thoroughly rinsed to remove all traces of alkali.
3. *Mechanical cleaning* includes wire brushing and groove and rotary roughening.
4. *Flame cleaning* makes use of heat to drive off moisture and to remove grit and scale.
5. *Sandblasting* is used for the removal of oxide scale, particularly when a rough surface is desired. It is a very effective surface preparation method.
6. *Pickling and etching* employ immersion of the material in an acid bath to remove scale. Most metals can be pickled in an acid bath—aluminum is the exception. Articles pickled and etched must be rinsed thoroughly to remove all traces of acid.

170

Metal Coatings

Metallic coatings can be divided into two types: anodic and cathodic coatings. The metal to be protected is called the base metal, whereas the metal used for protection is called the coating metal.

1. *Anodic coatings* are produced from metals which are anodic to the base. For instance, zinc, aluminum, and cadium coatings are anodic on steel because their solution potentials are greater than that of the base metal. The anodic coating protects the base metal by dissolving into solution with it.
2. *Cathodic coatings* are obtained by the application of metal more noble than the base. They protect the base because they have higher corrosion resistance. However, if pores or cracks are present in the coating, more severe damage could result than if the base had no coating at all.

Metallic coatings can be applied by electrodeposition, chemical immersion, spraying, cementation, chemical conversion, hot dipping, or by vapor deposition. Other means of protection are with organic coatings or porcelain enamels.

Electrodeposition   In electrodeposition or electroplating, a metal is deposited on the base metal by passing a direct current through an electrolyte solution containing a salt of the coating metal. It is one of the most important methods for the commercial production of a metallic coating. The electroplated metals used most frequently are zinc, aluminum, chromium, tin, and copper. Alloys are being increasingly used in electroplating.

Chemical Immersion   In this process, the base metal is immersed into an electrolyte solution containing a salt of the coating metal. Deposition takes place either by simple displacement, where the ions of the nobler metal are displaced from the salt solution by atoms of a less noble metal, or by reduction of an existing surface coating in which the base metal does not react.

Immersion coatings are very uniform and are usually thin. They are often used to prepare surfaces for other metallic or organic protective coatings. Nickel, tin, gold, and silver coatings are often produced by immersion. Zinc is deposited on aluminum or magnesium as a base for further nickel plating.

The immersion plating process is also used for nickel and results in a strongly adhered coating of considerable hardness. As the coating is 'thick', it is nonporous and exhibits a corrosion resistance equivalent to nickel-clad steel. High wear resistance makes the coating useful for instrument gearing. The coatings are also widely used on tank interiors to prevent corrosion/contamination from caustic soda, ethylene oxide, and tetraethyl lead. Other applications include gas storage bottles for liquid fuel rockets.

Tin immersion coatings are noted for their low cost, low frictional properties, and ease of application. However, their corrosion resistance is only fair. Tin coatings are applied to aluminum pistons to provide lubrication during 'running in' periods. Gold coatings are cheap; however, they tarnish and are only used for minor electrical parts and inexpensive jewellery.

Metal Spraying   Sprayed or metallized coatings are obtained by impinging molten coating particles against the base metal. When the molten particles strike the metal surface, they flatten into flakes and interlock with surface irregularities. A variety of guns have been developed for spraying molten metal. The designs depend upon the form of the metal when introduced into the gun, the kind of material being sprayed, and the temperature which is required.

Almost any metal, alloy, or ceramic material may be sprayed. The coating material is melted by oxyacetylene in metallizing and powder spraying, or by ionization as a gas in an electric arc. The sprayed metal structure consists of laminated flakes conforming to the metal surface, producing continuous, although somewhat porous, coatings. Advantages of metal spraying include speed of working over large areas and ease of application of thick coatings. Sprayed deposits are porous, which is an advantage for bearing surfaces. However, corrosion may be a problem with porous coatings that are anodic to the base material unless paints for sealants are used.

Cementation or Diffusion   The base metal is heated in the cementation process with a powdered coating metal to a temperature sufficient to allow for diffusion of fine particles. This results in the formation of layers of varying composition. The layer adjacent to the base metal surface may be an intermetallic compound or a solid solution, or diffusion may take place at grain boundaries only. The outer layers are richer in the coating material. Coating thickness is controlled by varying time of treatment and temperature.

The coatings are known by the following means: sheradized (zinc coatings), calorized (aluminum coatings), and chromized (chrome coating). Siliconized coatings are obtained by the application of silicon on molybdenum to protect it against high temperature oxidation. Cementation is also used for coating iron and steel with beryllium, boron, manganese, tungsten, vanadium, and zirconium and in the manufacture of threaded parts, bolts, screws, valves, and gage tools.

Chemical Conversion Coatings   Chemical conversion coatings are inorganic barriers produced by chemical or electrochemical reactions with the base metal surface. They differ from paints and most metallic coatings in that they are an integral part of the base metal. They are particularly useful as a base for covering with paints, enamels, and lacquers. Conversion coatings may be divided into phosphate, chromate, chemical oxide, and anodic coatings.

Phosphate coatings are produced by the chemical reaction of phosphates or phosphoric acid with the base metal. They are applied mainly to iron, steel, and zinc by immersion, spraying, or brushing. They do not provide much protection and are used generally as a base for painting. Although coatings can be made quite thick, they are easily damaged. Chromate coatings are used for the protection of zinc, aluminum, and magnesium parts and are applied by solution immersion. The nonporous film is more corrosion resistant than phosphate coating, but has low abrasion resistance. Chromating is used for aluminum parts in aircraft. Chemical oxide coatings are produced by treating the base metal with an alkaline oxidizing agent to provide bases for paints, oils and lacquers. Anodizing is the process of forming a thick oxide surface coating and uses the base metal as the anode of an electrolytic cell. Such coatings may be up to 1000 times as thick as natural oxide coatings. Anodizing is used for aluminum, magnesium, and their alloys. The electrolytes used are chromic and phosphoric acids for aluminum and sodium dichromate for magnesium. Anodic coatings are used on pistons, piston rings, cylinder liners, connecting rods, helicopter blades, and refrigerator shelves.

Organic Coatings

Organic coatings are made of inert materials and provide corrosion protection and decorative features. The protective value of an organic coating depends on its chemical inertness to corrosive environments, good surface adhesion, and impermeability to moisture and gases. Impermeability depends on the coating thickness. Thicknesses greater than about 0.4 mm are referred to as mastics or linings rather than as coatings. Organic coatings include, paints, enamels, varnishes, and lacquers.

Paint is a mixture of an oil and a pigment. It protects against moisture and weathering. Several agents can be added to achieve quick drying and reduced viscosity. Paints are most economically applied to plant equipment and structures with readily accessible surfaces. Varnishes consist of a drying oil, a resin, and a solvent thinner. On application to a metal surface, the thinner evaporates and the oil/resin mixture oxidizes and polymerizes to form a clear dry film. Varnishes are used where exterior durability in a clear finish is desired. Hardness and other properties can be varied over a wide range. Hard varnishes are used in aluminum paints. Lacquers are solutions of cellulose derivatives which may contain resins to contribute film hardness and plasticizers to provide flexibility. Solvent resistance of lacquers can be obtained by incorporating alkyds.

An enamel consists of a dispersion of pigments in a varnish or resin vehicle. It differs from varnish in being pigmented. The properties of enamels vary widely, depending on the varnish vehicle and resins. Drying can take place in air or at elevated temperatures. Shellac, emulsions, organosols, and bitumenous coatings are also used. Organosols are coatings in which a resin, usually PVC, is suspended rather than dissolved in an organic fluid. Organosols are tough and moisture resistant coatings. Although they have good electrical resistance, they are frequently used as secondary insulation to reduce shock hazards. Exposure to temperatures greater than 100 °C causes degradation. Organosols are used principally in car interiors. Bitumenous coatings are mistures of gaseous, liquid, or solid hydrocarbons. They have good abrasion resistance and insulation properties and are used in buried pipelines and for car body undersealing.

Porcelain Enamels

Porcelain enamels are vitreous coatings used to provide protection against acids and chemicals, and abrasion and wear resistance. Enamels usually have high temperature resistance as well. Mechanical properties of porcelain enamels are comparable with organic coatings and they are fairly damage resistant. Often a thin coating is desired to minimize impact or thermal shock damage. Thick coatings are an advantage in many chemical environments.

References

BERRY, R.W., HALL, P.M. and HARRIS, M.T. *Thin Film Technology*, Van Nostrand, New York, 1968

BEGEMAN, M.L. and AMSTEAD, B.H. *Manufacturing Processes*, John Wiley, New York, 1968

Mechanical Finishes and Facings

This section describes mechanical surface finishes, machining processes as they affect surface finishes, and facings which are used to harden surface finishes.

Finishes

A frequent problem facing design engineers is specification of surface finish. Quality of the finish on running or sliding parts greatly influences product life. Additionally, a smoother finish than required is a waste of machine time and hence money. The term 'surface finish' is usually applied to cold worked metals. For hot worked metals, the surface roughness that results from the deformation process is usually small in comparison with oxidation effects.

A major characteristic of a surface is *roughness*, which refers to finely spaced irregularities, the height, width, and direction of which establish the predominant pattern. Surface irregularities on lathes are produced by the cutting tool and machine feed. The height is found using an arithmetical average. Roughness itself does not alter the alignment of a surface. Another type of surface irregularity is *waviness*, which is due to irregularities of greater spacing than roughness. They may result from machine or work deflection, vibration, or heat treatment. The height is defined as the peak-to-valley distance and the width is the spacing of adjacent waves. Another surface characteristic is the *lay*, which is the predominant direction of the surface pattern. Lay is determined normally from the method by which the material was worked. *Flaws* are irregularities that occur at a location or at relatively infrequent intervals on the surface—e.g., scratches, ridges, holes, peaks, or cracks are flaws.

Quantitative values can be given to surface finishes by using a mean line through the profile as a datum. The common standards are the center line average and the root mean square. Australia uses the center line average, which is defined as the arithmetical average value of the departure of the profile, both above and below its mean line over a prescribed length.

Machining Processes

Machining processes can be divided into two groups: primary machining methods, which use single- or multiple-edged tools and remove large amounts of metal in heavy roughing cuts, and secondary methods, which usually follow

174

primary machining and impart greater dimensional accuracy, improved surface qualities, and less residual stress.

The two main primary machining methods are turning and milling. There are several sources of roughness when machining on a lathe with a single point tool. Feed marks are left by the cutting tool and built-up edge fragments may be embedded in the surface during chip formation. Chatter marks may occur from vibration of the tool, workpiece, or the machine. The feed causes a roughness which lies in the axial direction; gouging and chatter marks are also cyclic because the workpiece is revolving. Milling operations, in which the workpiece is stationary while the tools moves, produce surface finishes similar to lathe operations.

Secondary machining methods include grinding, lapping, honing, and superfinishing. Grinding can be thought of as milling with hundreds of small cutters, being essentially a scratching action. Types include cylindrical grinding, surface grinding, and centerless grinding. Roughnesses of the order of 0.01 mm are possible.

Lapping, honing, and superfinishing are used to obtain very smooth finishes. Lapping is an abrading process performed with either abrasive compounds or solid bonded abrasives. By lapping, gage blocks are finished to $\pm 50 \times 10^{-6}$ mm per millimetre of length. The reason why lapping is more accurate than grinding and some other finishing operations is that very little heat and pressure (which induce strains in the finished part) are involved.

Honing deals primarily with internal surfaces. Fine abrasive particles are used for the cutting action. In honing, very little heat is generated and there is no submicroscopic damage to the workpiece surface. Tolerances are easily maintained within 0.0025 mm. A crosshatch pattern of varying angles may be imparted to the work surface to enhance lubrication-holding qualities.

Superfinishing is somewhat similar to honing, but is applied primarily to external surfaces. Like honing, it also uses an abrasive stone. Unlike honing, superfinishing is not designed to remove material or correct part geometry. The amount of material removed may range from 0.0025 to 0.01 mm. The surface finish produced can be as low as 30×10^{-6}mm.

Another finishing operation is polishing. Polishing is the removal of tool marks on a part by the use of abrasive wheels and belts where accuracy need not be controlled.

Other rough processes include rolling and tumbling which remove burrs and scale, and improve surface finish. The process consists of placing a number of workpieces in a six- or eight-sided barrel with an abrasive medium. The barrel is rotated so that superfluous stock can be removed by the abrading action of the medium used. Another machining operation in which surface finish is not usually critical is drilling. In practice, the sides of the hole tend to be burnished by the rubbing action of the drill flutes. In special parts, drilling may be followed by reaming or honing to achieve a smoother surface finish.

Facings

Facings are hard surfaces applied to metals to give the important combination

175

of a hard-wearing exterior along with a tough, ductile core. They can be divided into two general categories: surface hardening and hard-facing.

Surface Hardening   Surface hardening techniques are employed to provide hard, strong, wear resistant surfaces on material that would be impossible to harden throughout or on parts that are not otherwise conveniently heat treat d. Surface hardening treatments are also used to provide surface properties that would not be suitable for the entire part.

There are two types of surface hardening treatment: the first involves local heating and quenching to obtain improved surface strength and hardness without altering core properties; the second type involves altering the surface composition with or without subsequent heat treatment of the entire part. Processes which come under the first type are flame hardening and induction hardening.

Flame hardening is used on small lots because of its adaptability and because no tooling is required. This method involves rapid heating of the part by means of a torch or torches followed by quenching. The part may be moved under the nozzles, or the nozzles may move over the part. The shape and position of the nozzles are modified to suit the shape of parts, such as gear teeth or flat or V-shaped lathe beds. With flame hardening, surface cracks may occur in quenching of high carbon steels, and the quenching rates must then be reduced. Cast iron lathe and machine toolways are flame hardened to give a hard wearing surface. Little distortion occurs, consequently little machine finishing is required.

Induction hardening uses a heating circuit that is essentially a transformer, where the conductor, carrying a high frequency alternating current, is the primary and the material to be heated is the secondary. The current flow sets up a magnetic flux of a circular pattern which passes through the surface of the work. A flow of energy is therefore induced and internal molecular friction developed in the work generates a hysteresis effect. When various parts are to be hardened, the induction coils must be changed to accomodate the particular part geometry. Induction hardening is a fast and reliable surface hardening method, however, it is usually economical only for production work.

Carburizing—which alters surface composition—is another surface hardening process, variations being pack, gas, and cyaniding. Carburizing involves heating a steel part to between 900 and 1000 °C in the presence of either a carbonaceous material such as charcoal, a carbon-rich gaseous atmosphere, or a liquid salt. The process is usually applied to low carbon steels. Open-fired, semi-muffle, electric, and gas-fired furnaces are used for controlled atmosphere carburizing. In all carburizing processes, the carbon content and depth of surface (or case) hardening are determined by temperature and time. The greatest amount of control and greatest case depths are obtained with gas-carburizing. Cases between 7.6 and 10 mm deep are obtained readily by gas-diffusion, while case depths to about 1.8 mm are obtained by pack carburizing. The salt bath method is particularly adapted to small parts where many components may be immersed and treated simultaneously. Case depths to 0.8 mm may be obtained by this method.

Cyaniding is performed on steels at temperatures between 820 and 900 °C in either salt baths or in electric or radiant-tube furnaces. The cyaniding technique provides a case of high hardness and good wear resistance. The process is the result of iron nitride formation in the case brought about by the release of nitrogen from the bath. It is used especially on smaller parts that can be treated without mechanical handling equipment.

Nitriding subjects parts to be hardened to ammonia gas at temperatures of $490 - 530$ °C, introducing nitrogen to the surfaces. Since those temperatures are usually below the critical hardening temperature of most steels, it is said to be a 'subcritical' hardening operation. Sealed retort furnaces with close temperature control are required for the process. The time of exposure is relatively long, 72 h being required for about 0.65 mm case depth. No quenching is necessary. The nitrided surface hardnesses range from 60 to 72 Rockwell C and case depths from 0.05 to 0.9 mm are obtained. Case depths in all instances should be no greater than necessary for reasons of economy and to avoid cracking of parts subjected to elevated temperatures.

Hard Facing   Hard facing is a process of applying better quality material to a part where wear is likely to occur, or has already taken place, in order to improve and extend surface life. The two processes used for application of hard facings are welding and flame spraying. Practically any welding method, such as oxyacetylene or arc, may be adapted. Hard facing materials include tungsten carbide composites, chromium carbide, austenitic and martensitic irons, cobalt base alloys, nickel base alloys, and austenitic, martensitic, and pearlitic steels. An example of hard facing usage is in coal pulverizers. Pulverizer blades are often hard faced with tungsten carbide to prevent erosion by coal particles.

In flame spraying, metal or alloy powder is sprayed through a reducing flame on to the section to be faced and is fused at about 1000 °C. Two types of powder in common use are self-fluxing alloy powders and carbide powders. Self-fluxing alloy powders are usually nickel based and cobalt based alloys containing chromium, copper, molybdenum, tungsten, carbon, and iron. Boron or carbon also contained in the powder determines hardness of the alloys, while boron combined with silica forms a thin protective borosilicate 'glass' slag on the fused surface. Carbide powders are usually mixtures of carbides (generally tungsten carbide eutectic) and self-fluxing alloys. Hardnesses vary from 30 to 65 Rockwell C; however, hardness alone does not guarantee abrasion resistance, which is a function of hardness and toughness. A correctly fused flame-sprayed self-fluxing alloy (having a hardness of about 35 Rockwell C) will often outwear a steel hardened to 60 Rockwell C, while hard faced coatings (having a hardness of 50 Rockwell C) will outwear the same steel by a factor of ten or more.

Rust Prevention

Rust is the product of corrosion of ferrous metals. Corrosion is defined as the destruction or deterioration of a material because of reaction with its environment. It is usually an electrochemical process.

An electrode polarized to a potential more noble than its reversible single potential acts as an anode and generates electrons, while another electrode polarized to a more base potential acts as a cathode and consumes generated electrons. Thus, if two different electrodes are connected they will adopt a common mixed potential and the more noble electrode (the anode) will deteriorate or corrode.

It might be thought that a pure homogeneous metal would not be affected by electrochemical corrosion; however, in pure metals there are grains and grain boundaries at different potentials. This is even more pronounced in some alloys. Corrosion may also be caused by an ion of different potential dissolved in the electrolyte.

The control of corrosion of ferrous metals comes under four headings: control of environment, design, electrochemical control, and control by coatings. Control of the environment includes reducing the humidity, addition of chemical inhibitors, control of the degree of aeration, pH control, bacteria control, and temperature control.

Control of Environment

Humidity control prevents the formation of water vapor and thus eliminates the electrolyte for a 'possible' electrochemical cell. For most common materials, a relative humidity below 30% will maintain corrosion rates at negligible values. If humidity control is supplemented by the use of protecting oils and greases, the critical humidity may be as high as 45%. However, the presence of dirt and rust particles may cause corrosion at humidities as low as 20%.

Inhibitor additives These are of several types, depending on whether they influence the anode, cathode, or both electrode reactions. They act as chemically or physically adsorbed films on the metal, and either alter electrochemical characteristics or serve as mechanical barriers to normal corrosion processes. Care must be exercised when using additives, as in some instances they may actually promote corrosion reactions.

Degree of aeration In neutral and alkaline conditions, the cathode reaction is reduction of oxygen to hydroxyl ions. By elimination of oxygen, the cathode reaction will not proceed, which in turn prevents the dissolution process at the anode from taking place. Control of oxygen in boiler feed water is an example. Certain materials such as stainless steel actually require the presence of oxygen to maintain the passive conditions, so in such circumstances aeration should be considered rather than deaeration for corrosion control.

pH control By maintaining neutral of alkaline conditions, the hydrogen-type cathode reaction is suppressed, which in turn reduces the rate of anode reaction.

Bacteria control In bacteria control, aeration of backfill and application of germicides reduce the effects of anaerobic bacteria which aggravate corrosion of buried pipelines and other vessels such as fuel tanks.

Temperature control This entails avoiding hot spots and fluctuations beyond

178

design temperatures, as increasing temperature generally increases corrosion rates.

Control by Design

Variations in temperature, electrolyte flow, velocity and concentration, and aeration of a solution, represent common sources of corrosion difficulties. Crevices, deposits, loose or discontinuous scales, sharp corners, obstacles to flow, and other causes of localized turbulence should be avoided or removed as they develop. Galvanic action as the result of poor design is a common source of corrosion problems. Very often it is necessary to use dissimilar materials, but under such circumstances several basic principles should be followed:

1. Avoid unfavorable area effects.
2. Insulate dissimilar metals, i.e., increase electrical resistance of the metallic conductor between anode and cathode.
3. Apply coatings either on the more noble metal (the cathode) or on both, but never on the anode alone.
4. Design for ready replacement of anode parts or make them of heavier sections.
5. Avoid threaded sections.

Electrochemical Control

Electrochemical control includes both anodic and cathodic protection. Anodic protection recognizes that many materials exhibit passive behavior under specific conditions. An example is the use of mild steel equipment to handle sulfuric acid. In most cases, however, the more common protection is cathodic protection. It is a well established principle and may be of two types: impressed current protection and sacrificial protection.

Different reactions occur at both the anode and cathode. The anode reaction produces electrons and the cathode reaction absorbs electrons. Corrosion cannot take place where the cathode reaction is proceeding as corrosion always occurs at the anode and not at the cathode. In cathodic protection, a sacrificial anode may be used to form a galvanic couple. It is selected so that it corrodes and not the more noble material being protected. Typical materials are zinc, aluminum, and magnesium. When these sacrificial material have corroded, they can be replaced easily and the protection continued. Alternatively, an impressed e.m.f. may be used for cathodic protection. Table 12 lists the galvanic series of metals and alloys for sea water conditions. Generally, the further apart are two materials in the series, the greater is the corrosion at the anodic end.

Control by Coatings

Coatings cover a wide range of materials, including metals that act by electro-chemical control, e.g., zinc, nickel, and chromium, and non metallic coatings (paint, varnish, lacquer, plastics and rubber) which act as barriers between the metallic material and its environment. Effectiveness of the barrier depends on the porosity and permeability of the coating to moisture and on the bond between the metal and its coating.

In metal coating, it is usual to apply 'printing coats' directly on to the metal.

Table 12   Galvanic Series of Metals and Alloys (Sea Water)

Magnesium and its alloys	Corroded end—anodic or
Zinc	least noble
Aluminum 2S	
Cadmium	
Aluminum 24S-T	
Steel or iron	
Cast iron	
13% Cr – iron (active)	
Ni-Resist	
18 – 8 stainless	
18 – 8 – 3Mo steel	
Hastelloy C (active)	
Lead – tin solders	
Lead	
Tin	
Nickel (active)	
Inconel (active)	
Hastelloy A	
Hastelloy B	
Chlorimet 2	
Brasses (Cu – Zn)	
Copper	
Bronzes (Cu – Sn)	
Copper – nickel alloys	
Monel	
Silver solder	
Nickel (passive)	
Inconel (passive)	
13% Cr – iron (passive)	
18 – 8 stainless	
18 – 8 – 3Mo steel	
Hastelloy C (passive)	
Chlorimet 3	
Silver	
Titanium	
Graphite	
Gold	Protected end—cathodic
Platinum	or most noble

These normally contain inhibitive pigments such as zinc chromate and zinc dust to prevent any action should moisture ingress occur. 'Top coats' are of various types, and selection depends on the final coating thickness required and the service under which they operate. Examples include organic zinc coatings (so called because of their binder type), epoxy, chlorinated rubber, and polyethers. They are typified by high zinc loadings (up to 95% zinc dust in the dry film). Performance of such coatings depends on the binder and application method.

When using nonmetallic coatings, the metal surface must be properly prepared and the paint system applied properly. Surface preparation involves surface roughening to obtain mechanical bonding as well as the removal of dirt, rust, mill scale, oil, and grease. In humid environments the interval between surface preparation and coating application is important, since a clean metal surface is highly susceptible to tarnish in a short time. The presence of films reduces bond strength and may act as preferential corrosion sites. An example of a nonmetallic coating is polyurethane, which forms an extremely hard, flexible, and abrasion resistant coating. Polyurethane also has a high degree of resistance to solvents, oils, and many chemicals and exhibits very high gloss and gloss retention, thus providing a good combination of protection and decoration.

References

'Corrosion, Erosion, and Lubrication Symposium', Institution of Engineers Australia, July 1971

DeGarmo, E.P. *Materials and Processes in Manufacturing*, 3rd edn, Macmillan, London, 1970

Lindberg, A. *Processes and Materials of Manufacture*, Allyn and Bacon Inc., Boston, 1969

Nickel, B. and Draper, A. *Product Design and Process Engineering*, McGraw-Hill, New York, 1974

Bearing Materials

Before describing various bearing materials, the different types of bearings available and their uses are outlined briefly.

Bearings

The term *antifriction bearing* is used to describe that class of bearing in which the main load is transferred through elements in rolling contact rather than in sliding contact. In a rolling bearing the starting friction and the running friction are about the same and the effects of load, speed, and temperature variation on friction are small. It is somewhat of a misnomer to describe a rolling bearing as 'antifriction' since some friction exists. In design and selection of antifriction bearings, factors such as fatigue loading, friction, heat, corrosion resistance, lubrication, machining tolerances, assembly, use and cost must be considered.

In a 'sleeve' bearing, a shaft or 'journal' rotates or oscillates within a sleeve or bearing and the relative motion is sliding. Applications of journal bearings are many. If, for example, an easily installed bearing with little or no lubrication is required, an antifriction bearing might be a poor selection because of cost, elaborate enclosures, close tolerances, radial space required, high speeds, or large inertial effects. Instead, a nylon journal bearing requiring no lubrication, a powder metallurgy bearing with the lubrication 'built in', or a bronze bearing with ring-oiled, wick-feed, solid lubricant film, or grease lubrication might be a very satisfactory solution.

Many ball bearing manufacturers use steels of the same composition (E52100 steels) and subject them to about the same heat treatment. The cages of rolling bearings are divided into pressed and machined types with different shapes according to bearing type and use conditions. For some bearing cages, synthetic resins are used.

Metallic Bearing Materials

Five distinct forms of lubrication may be identified:

1. hydrodynamic—thick liquid film;
2. hydrostatic—air or water, pressurized;
3. elastohydrodynamic—mating gears, rolling bearings;
4. boundary—only several molecules thick;
5. solid film—graphite or molybdenum disulfide.

The above lubrication processes affect selection of bearing materials. Several conflicting requirements of a good bearing material are that it must have satisfactory compressive and fatigue strength to resist externally applied loads and, at the same time, have a low melting point and a low modulus of elasticity. The latter requirements are necessary to permit the material to wear or 'break in', since the material must conform to slight irregularities and absorb and release foreign particles. Resistance to wear and low coefficient of friction are also important because all bearings must operate, at least for some time, with boundary lubrication only. Additional considerations in the selection of bearing materials are ability to resist corrosion and cost.

Small bushings and thrust collars are often expected to run with thin-film lubrication. In such cases, improvements over solid bearing materials can be achieved with porous materials. Bushings made by powder metallurgy techniques are porous and permit oil penetration. Sometimes porous bushings are enclosed by oil-soaked material. Bearings are frequently ball-indented to provide small basins for the lubricant storage while the journal is at rest, thus supplying some lubricant during starting. Another method of reducing friction is to indent the bearing wall and fill the indentations with graphite. Although such bearings do not 'bed down' as well as a thick babbitt layer, three-layer composite bearings operate at very high loads and are used frequently in heavy duty automotive, truck, diesel, and aircraft engines.

Aluminum Alloys Aluminum alloy bearings offer excellent resistance to

182

corrosion by acidic oils, good load-carrying capacity, superior fatigue resistance, and good thermal conductivity. Poor scoring characteristics are improved by alloying with tin, lead, or cadmium, but at the expense of reduced bearing strength. For this reason, a steel backing is often used with a thin layer of bonded aluminum. A lead babbitt overlay gives excellent ductility and scoring resistance. Aluminum alloys are used generally in heavy duty bearings in tractors, diesel, and aviation engines. Not all aluminum alloy applications have been successful—bearings under omnidirectional loads have not been as satisfactory as with reciprocating loads.

Babbitts Babbitts, also called white metal alloys, are divided into tin base and lead base alloys. They represent the majority of sliding bearing alloys used. With use of babbitts, there is a minimum tendency for damage to steel journals under conditions of boundary lubrication or in dirty operation; they have excellent compatibility and nonscoring characteristics, as they are easily extruded and smeared during sliding. They are outstanding in embedding dirt and in conforming to geometric variations in machine construction, due to the soft matrix. A typical composition of a tin based babbitt is 7.5% antimony, 4.5% copper, 0.35% lead, and the balance tin; that for a typical lead based alloy is 10% antimony, 5% tin, 0.5% copper, and the balance lead.

Tin based babbitts are more desirable generally than lead based types because of better corrosion resistance. They function well under conditions of poor lubrication, and provide easier bonding to a steel shell. The high cost of tin and occasional shortages, however, have resulted in wide use of lead babbitts.

Bronzes Bearing bronzes may be grouped into three categories: lead bronzes, tin bronzes, and high strength bronzes. These have better high temperature properties than babbitts but poorer compatibility, conformability, and embeddability properties—particularly with low lead content bronzes. Hence it is advisable generally to use the softest bronze possible, while still having necessary strength and load carrying capacity. Bronzes are used in large volume for cast bushings because they offer adequate bearing properties, have excellent casting and machining characteristics, are relatively hard and strong and hence do not require a steel backing, and are low in cost.

Cadmium Cadmium alloys are used only in special applications because of their fairly high cost. The bearings have excellent fatigue life, but are attacked by corrosive oils unless treated with indium.

Carbides Cemented carbide bearings consist of carbide in a cobalt matrix, forming a very wear resistant material having excellent surface finish. This type of bearing can carry higher loads and speeds than practically any other known material. Use is limited by cost and the difficult manufacturing processes required.

Copper – Lead Alloys Simplest of the many copper-base materials is the binary system containing from about 20 to 40% lead. As lead is practically insoluble in copper, the alloy consists of a copper matrix containing dispersed lead particles. Like babbitts, copper – lead alloys owe their frictional properties to the spreading of thin films of soft lead over the harder copper surface. Silver is sometimes added to minimize segregation with high lead contents. Corrosion can be a problem with these alloys, hence indium may be diffused into the bearing surface. The alloys are manufactured by a coating or sintering process using powdered metals.

Hardness of copper – lead alloys is not better than that of babbitts at room temperature but it is better at temperatures of 150 °C and above. Although strengths are greater than those of babbitts, steel backings are usually used for increased strength. High load-carrying ability is obtained with the use of copper – lead – tin bearing materials.

Silver Silver bearings have found extensive use in heavy duty applications in the aircraft industry. They normally consist of electrodeposited silver on steel backings with an overlay of 0.025 – 0.13 mm of lead. Indium is usually flashed on top of the lead overlay for corrosion protection. Such bearings have outstanding metallurgical uniformity, excellent fatigue resistance and thermal conductivity, high load carrying capacity and can operate at high temperatures. High cost is a major disadvantage in their use.

Zinc Zinc castings are low cost and are used where bearing speeds and loads are low. Sintered powder-metal bearings are known as 'oil-less' or 'self-lubricating' bearings. They are made having controlled voids, so that they can contain 15 – 35% impregnated oil. They are, therefore, both bearing and lubricant and, if properly made, function for life in many applications. Porous bronzes containing 90% copper and 10% tin are the most suitable, but other compositions such as iron base alloys are also used.

Other Materials

Many nonmetallic materials are available for bearing applications, the most common being plastics. Rubber has also found specialized uses in the drive shafts of ships and in other devices where water is present and a bearing material is required which will embed dirt and other foreign material. Nonmetallic bearings and sintered metals are used frequently in the same type of applications. The excellent frictional and wear characteristics of some plastics, which often enable operation with little or no lubrication, have been major factors in their expanding use. Low cost, light weight, resiliency, quiet operation and corrosion resistance have also contributed to their usage.

Composite bearings are made of filled plastics and porous metal supporting structure to provide strength, dimensional stability, and good heat transfer. Although many plastics have found at least limited use as bearing materials, those which are used most widely are the injection molding plastics Lexan and Delrin, laminated phenolics, nylon, and Teflon.

Injection Molding Plastics Two common plastic bearings materials are the polycarbonate 'Lexan' and the acetal 'Delrin'. Since these materials can be formed readily, cost savings over phenolic laminates or metallic bearings are often possible.

Lexan offers exceptional impact strength and is highly resistant to creep and cold flow. It is also transparent, has good wear resistance, and is used as a structural bearing material where transparency is required. It finds use in rollers and bushings at temperature to 120 °C. Delrin has good creep resistance and other mechanical properties, and excellent solvent resistance. Delrin is used in bearings for electrical switches, electrical shavers, egg-beaters, pulleys, lawn mower rollers, and door hinges.

Laminated Phenolics For various heavy duty industrial and marine applications, excellent bearing service can be obtained with phenolic plastics using cotton, rayon, asbestos, or other fillers. Up to 10% graphite is sometimes added to enhance self-lubricating properties. Excellent strength and good abrasion and wear resistance are offered by phenolic bearings. They also provide high impact strength, low friction, good lubrication characteristics, resistance to deformation, and good machinability.

Nylon In small lightly-loaded bushings, low friction nylon requires no lubrication. It is easily molded to close tolerances, is wear and abrasion resistant, and is quiet in operation. Fillers such as graphite and molybdenum disulfide can be used to improve resistance to cold flow, mechanical properties, and wear resistance.

Teflon Teflon is unique among plastics because of its unusually low coefficient of friction, wide service temperature range, resistance to attack by chemicals and solvents, and ability to operate without lubrication in many applications. Despite these advantages, high cost and low load capacity have lead to its use in modified form—often with inexpensive fillers or reinforcing materials such as glass fiber.

References

SHIGLEY, J.E. *Mechanical Engineering Design*, McGraw-Hill, New York, 1963
BAER, E. *Engineering Design for Plastics*, Reinhold, New York, 1964
WILCOCK, D.F. and BOOSER, E.R. *Bearing Design and Applications*, McGraw Hill, New York, 1975
LANG, J.E. *Bearings in Structural Engineering*, Newnes-Butterworth, London, 1975

Lubricating Oils

Lubricating oils form a vast family of fluids which are considered universally synonymous with the moving parts of all types of machinery. The first lubricants were animal fats, although these were later replaced by petroleum fractions. Modern lubricating oils are of many types and grades, ranging from heavy petroleum gear oils to high speed synthetic oils for turbines. The main functions of lubricating oils are:

1. to reduce wear between moving parts;
2. to reduce friction and thereby save power;
3. to dissipate heat;
4. to protect against corrosion;
5. to reduce noise, vibration, and shock between parts such as gear teeth;
6. to flush away contaminants;
7. to act as a carrier for additives.

The different types of oils, their production, properties, applications and additives are discussed in this section. First, however, the mechanisms by which lubricating oils achieve their purpose are mentioned. The two mechanisms used by oils to effect lubrication are fluid lubrication and boundary lubrication. Fluid lubrication, also called hydrodynamic lubrication, involves a complete separation of the moving surfaces by a thick layer of fluid lubricant, so that there is no direct contact. The effective coefficient of friction becomes very low and is a function of fluid viscosity. For this reason, and because wear is zero, fluid lubrication is desirable. In practice, however, it is often impossible to maintain a continuous fluid film between bearing surfaces as starting and stopping usually causes film breakdown, as may vibration and excessive loading in service. The mechanism of boundary lubrication occurs when the oil layer is only a few molecules thick, comparable to the surface roughness of the moving surfaces. Boundary lubricants are produced by adding a small quantity of polar organic compounds as fatty acids to a straight lubrication oil. A monomolecular layer strongly adherent to the lubricated surface forms from the reaction of polar groups and the surface. Another fluid layer forms on top of the first, and the process continues to a maximum layer thickness of about 100 nm. The final film has high compression resistance but low shear resistance, giving a very low coefficient of friction. In many instances, the actual mechanism is a hybrid of boundary and full-fluid lubrication, due to the impossibility of maintaining complete separation of the parts.

Types of Oil

Lubricating oils can be divided into five major categories: mineral, asphaltic, fatty, fluid polymers, and synthetic oils. Although each category contains excellent lubrication oils, many commercial products are blends of two or more oils which have a more desirable set of properties than an oil by itself.

Mineral Oil Mineral or petroleum lubricating oils are the largest and most widely used family of oils. They are derived from crude oils by fractional or vacuum distillation, followed by de-waxing, solventing, de-ashing, and finishing. Not all crudes process to high quality oils. Paraffinic crudes yield oils which have relatively low viscosity, density, and high oxidation resistance. Aromatic oils have high oxidation stability but exhibit rapid changes in viscosity with temperature, while napthenic crudes provide oils having a low pour point and

186

low oxidation resistance. Mineral oils may be classified by their properties and are referred to as neutral oils, bright stocks, cylinder stocks, and residual oils.

The properties necessary to describe mineral oils correctly include viscosity, viscosity index, specific gravity, flash point, fire point, pour point, carbon residue, color, and neutralization number. The most important properties are viscosity and viscosity index, since these are directly related to the coefficient of friction. The viscosities of mineral oils are given as SSU (Saybolt second universal) and are determined by a viscometer. Absolute values are often converted to kinematic viscosity. Viscosities vary from 80 SSU for light oils to 5000 SSU for heavy steam-refined oils, a very wide range.

The viscosity index is an empirical number used to describe the effect of change of temperature on oil viscosity. The viscosity index ranges from 0 to 100. The higher the value an oil possesses, the less its viscosity changes with temperature variation.

Specific gravities of mineral oils are less than unity, the range being from 0.85 to 0.95. The specific gravity of an oil may be used as an indicator of the type of crude from which it originated. Paraffinics generally have low densities whereas napthenics are high density oils.

The flash and fire points of an oil define the upper service temperature which can be expected from that oil. The flash point is the temperature at which sufficient oil vapor is present to form an inflammable mixture with air, while the fire point is the temperature to which an oil must be heated to burn continuously. For mineral oils flash points vary from 150 to 250 °C, and fire points vary from 170 to 350 °C. The pour point of an oil is an indicator of its lowest service temperature, which for mineral oils varies from 0 to −40 °C. Properties of carbon residue, color, and neutralization number are used additionally to provide a complete oil specification.

Asphaltic Oils   Asphaltic oils are highly viscous and consist almost exclusively of asphalt residue from steam or vacuum refining. They may also be produced by mixing heavy grade fluxes with black mineral oils. Their densities are less than that of water while their pour, flash, and fire points are commonly 25, 280, and 300 °C respectively. Their use is limited to heavy duty, low speed gear systems.

Fatty Oils   Fatty oils are of animal or vegetable origin. Their properties are highly variable, being governed by nature, and thus they offer no competition to mineral oils. Their use has declined to extinction in industrialized countries.

Fluid Polymers   New technology has developed fluid polymers as lubricants, both as additives and fluid fractions. The polymers in these liquids must be compatible with other ingredients, mainly mineral oils, and complete polymerization is required so that elevated service temperatures do not cause thickening. A popular polymer, used as a gear lubricant, is polyisobutylene, having about the same physical properties as crêpe rubber. It is dispersed by a rubber mill in a lubricating oil, resulting in a solution named 'Paratac'. Other polymers used are styrene – polyisobutylene, polyethylene, and resins recovered from mineral oils.

187

Synthetic Oils Synthetic lubricating oils 'came of age' during World War II with efforts to develop mineral oil substitutes. The first synthetics were polymers of aliphatic olefins and had characteristics similar to the scarce mineral oils. Today many more exist, but their most important functions are to provide characteristics beyond the range of mineral oils. Thus their main areas of application are in low temperature, high temperature, and corrosive environments.

The main families of synthetic lubricating oils are diesters, phosphates, polyglycols, and silicones. It is difficult to differentiate between them, however, as properties are similar and in many instances oils consist of combinations of them. All types have a high viscosity index and have at least the same load carrying capacity as minerals oils. All have low pour points and high flash points when compared to mineral oils. The diester and phosphate oils have pour points to -80 °C and flash points to 230 °C, while the polyglycols have corresponding values of -45 °C and 220 °C. Synthetic oils have the greatest continuous service temperature of all oils. For silicone fluids, the flash point is as high as 330 °C and the maximum service temperature is about 300 °C. Polyphenyl ether has a maximum recommended service temperature of 375 °C, and is said to have a useful life of 30 h at 500 °C. All the synthetic fluids are chemically unreactive in varying degrees, from mostly unreactive for phosphates to completely unreactive for silicone oils. With reference to load-carrying capacity, a polyalkylene – glycol blend has the least load carrying ability of all the lubricating oils of similar viscosity, but has special application in worm-gear speed reducers.

The major disadvantage of most synthetic oils is their cost, which starts at about \$5 per litre (in 1979). Some types also present difficulties in certain applications, e.g., silicone oils are unsatisfactory in lubricating steel/steel surfaces. Because of the impending end of crude oil supplies within the next half century, petroleum fuels and mineral oils will eventually be prohibitively expensive. Consequently much work is being conducted to develop synthetic oils, some types being halogenated hydrocarbons, phosphate esters, silicate esters, neopentyls, fluorocarbons, chlorinated silicones, mercaptals, and chlorinated aromatics.

Applications

Lubricating oils are used in an enormous number of applications but may be considered in two major fields: industrial and domestic. Industrial applications can be further divided into primary and secondary industry.

Primary industry, particularly in Australia, uses large quantities of lubricating oils in mining operations. Draglines, shovels, scrapers, tracks, and conveyor systems are all lubricated usually by EP (extreme pressure) heavy-duty mineral oils, typically between SAE 80 and 140. These oils provide protection from weather and abrasive particles in air caused by mining operations. In forestry and sawmilling, mineral oils are used to lubricate saws, winch drums, loaders, and haulers. Hydraulic power transmission units are common and make use of light mineral hydraulic oils such as SAE 20W. Lubricating oils are used in a vast variety of agricultural machinery including tractors, bulldozers, harvesters, and other vehicles.

Secondary industry uses huge quantities of lubricating oils:

Power Production   In both steam turbine and hydroelectric power generation, mineral turbine oils meeting strict specifications are required. Viscosities are about 400 SSU and typical oils are 2135 T-H and 2190 TEP. In nuclear power plants, preference is given to solid lubricants such as MoS_2, but certain radiation resistant synthetic oils have been developed. An example is polybenzenoid compound oil.

Heavy Industry   The steel industry is typical. Thousands of large and small gears require lubrication. Oils used are mineral or asphaltic oils and are usually heavy residuals, intermediates, and extreme pressure (EP) oils. Equipment in the cement industry such as agitators, conveyors, elevators, mixers, grinders, and pumps must be lubricated and mild EP and residual mineral oils are used. These oils also prevent wear from abrasive dust. In the paper industry, rugged lubricating conditions are experienced in debarking and chipping which require mineral oils such as SAE 140 EP

Light Industry   In light industry are the medical, textile, and printing operations. In the medical field, highly refined, light-colored turbine oils with viscosities about 300 SSU are used to lubricate mills, mixers, tablet machines, and filling machines. Mineral oils of low viscosity and with EP additives are used in the textile industries, where speeds are high, throw-off is undesirable, and shock loadings are common. In the printing industry, low viscosity mineral oils are utilized. Additionally, it is in light industry that the lubrication of nonferrous and nonmetallic parts such as gears is required. For nonferrous applications, mineral oils containing tallow are used mainly, but in certain applications synthetics are used to great advantage.

Transportation Industry   This industry covers cars, trucks, cranes, heavy machinery, and diesel locomotives. Lubricants are required in engines, transmissions, differentials, axles, bearings, and suspension systems. Mineral oils are used with viscosities varying from low to very high. Some machinery contains hydraulic transmission systems wherein synthetic oils such as the phosphates and esters are as important as mineral oils. Diesel locomotive traction gears are lubricated by residual oils which have viscosities as high as 2000 SSU and good adhesion.

Marine Industry   In the shipping industry, propeller shaft bearings, turbines, gear reduction sets, steering gears, cranks, windlasses, and winches must all be lubricated and simultaneously protected from marine corrosion. Mineral oils are used generally with viscosities ranging from 300 to 500 SSU in the engine room, to over 2000 SSU for deck winches. Specially blended mineral turbine oils are used in the drive section of some ships.

Aircraft/Aerospace   Synthetic lubricating oils find their largest usage in the aerospace industries. Although propeller-driven aircraft are still lubricated with heavy body mineral oils, lubrication in jet-engined aircraft is monopolized by synthetic oils to achieve service temperatures to 330 °C, to reduce evaporation loss, and to eliminate coking. Typical oils are diesters and polyphenol ethers. Military forces use high performance synthetic oils in their aircraft. Not only in aircraft engines are the synthetics used; they have major applications as well in airframe lubricants in the form of diesters. Missiles and space vehicles also require lubricating oils, synthetics being generally used.

The domestic usage is almost completely supplied by mineral oils. Washing machines, sewing machines, and many appliances are typical applications. Synthetic oils have specialist uses such as gun oils containing silicones or aliphatic napthas to provide protection from corrosion and wear.

Additives

Oil additives are substances which impart or enhance the desirable properties of a lubricant and eliminate or minimize deleterious properties. They are added only in the small amounts necessary and are expensive compared to oils. Various additives are as follows:

Antifoam Agents   During machine operation, many oils tend to foam and become useless. To prevent foaming, silicone fluids in concentrations of 0.001% may be added to oils. Other additives include glycerol monostearate, sorbitol esters, methyl salicylate, and metal sulfonates.

Antioxidants   The prevention of oxidation in mineral oils subjected to continuous service temperatures above 100 °C is achieved by addition of antioxidants. Typical compounds are amines, phenols or naphthalics, disulfides or thioethers, of compounds of phosphorous, selenium, and tellurium. Percentages added vary from 0.1 to 1.0%.

Rust Inhibitors   Rust inhibitors are added to oils used for lubrication of ferrous parts where water is present. Typical inhibitors are barium, calcium, or sodium sulfonates, fatty amines, and phosphoric acid; these are added in percentages varying from 0.02 to 2%.

Corrosion Inhibitors   These compounds prevent corrosion of bearing materials caused by contaminants. The inhibitors are generally organic compounds containing phosphorous and sulfur and are added in amounts from 0.2 to 3%.

Detergents   Detergent additives are dispersants containing oil compounds in solution which would otherwise tend to deposit. They also keep mating surfaces clean. They are used in automobile transmission oils to ensure shock-free operation. Detergents consist largely of phenates and sulfonates in the form of barium and calcium salts.

190

Dyes Dyes are added to light colored oils to aid identification and to give the oil a distinctive appearance.

Extreme Pressure Agents EP additives form thin surface films which prevent wear when bearing materials are subjected to extremely high loads. There are over 60 varieties which contain chlorine, phosphorous, sulfur, or other materials.

Metal Deactivators These compounds prevent metals such as bronze from going into solution and causing rapid oil breakdown. A typical deactivator is quinizarin.

Odor Control Agents When EP agents are added to oils, the resulting blend often has an unpleasant odor which may be overcome by adding an odor control agent such as pine oil or pine tar.

Pour Point Depressants These additives are used to lower the service temperature of oils and are either acryloids or wax-condensation products from napthalene or phenol.

References

BONER, C.J., *Gear and Transmission Lubricants*, Reinhold, New York, 1964
BOWDEN, F. and TAYLOR D., *Friction and Lubrication*, Methuen, London, 1971
Caltex Lubrication: a journal devoted to selection and use of lubricants
Journal of Lubrication Technology, American Society of Mechanical Engineers

Cutting Fluids

The primary objective of a machining operation is to produce a component economically within some dimensional tolerance and surface finish. A cutting fluid must be able to assist in achieving the objective and possess the following capabilities: (1) increase tool life, (2) improve surface finish, (3) reduce cutting forces and machine power consumption, (4) reduce workpiece distortion due to temperature rise in the cutting zone, and (5) facilitate removal of chips.

Action of Cutting Fluids

During the cutting operation, friction is developed between chip and tool face, workpiece and tool, workpiece and noncutting edges of tool, and between chips and guiding surfaces of the tool. If friction in any of these areas can be reduced, the result will be decreased tool forces, improved surface finishes, or both. Low shear strength films can be formed on the clean and highly reactive surfaces of cut metal and reduce sliding friction. Reduction of friction between chip and tool and workpiece and tool are achieved by this means, since high temperatures and pressures in the cutting zone do not permit use of hydro-dynamic or boundary layer lubrication between the workpiece and the noncutting edges of the tool, or between chips and guiding surfaces. A cutting fluid having good lubrication properties is helpful in reducing friction.

The principal properties of cutting fluids are grouped under either performance or service properties. Performance properties include: (1) workpiece finish, (2) workpiece dimensional stability, (3) tool life, and (4) economic production. Workpiece finish is most often a problem in relatively low speed machining because of a heavy built-up edge which forms on the tool, breaks off on to the workpiece, and gouges freshly cut metal, thus damaging finish. Dimensional stability of the workpiece is influenced by the cooling efficiency of the fluid. Cooling prevents excessive workpiece distortion due to heat of deformation and frictional heat. Tool life is of particular importance in high speed machining because high tool-point temperatures can seriously shorten tool life. Economic production requires use of fluids having properties that produce the greatest number of acceptable pieces at the lowest unit cost.

Service properties of cutting fluids are often the best criteria for judging their efficiency. Desirable properties are: (1) no corrosive effects on machine parts, (2) ease of handling and preparation, (3) detergent action, and (4) long service life and minimum cost. Undesirable service properties are rusting, foaming, smoking, fogging or misting, development of odors, toxic effects on the operator, and deterioration of paint or finish on machines.

Types of Cutting Fluid

There are four major types of cutting fluid: aqueous, oil types, chemical or synthetics, and gases.

Aqueous Fluids These include emulsions, suspensions, solutions, and other mixtures that are diluted with water. Aqueous fluids are generally the most efficient for cooling, and are thus widely used for high speed machining operations. Emulsifiers that are often used include sulfonates, resin soaps, and glycols. The fluids may contain fatty oils or acids, chemical wetting agents, water-softening agents, and germicides.

Oil Type Fluids These are petroleum based and consist of a diverse range of oil mixtures, additives, that provide lubrication, friction reduction, and protect the tool and work surfaces against welding, galling, metal pickup, seizure, and tool failure. Oil type fluids may be divided into two categories: inactive and active. Those that are 'inactive' (i.e., do not react chemically with the metal surface) include mineral oils, fatty oils or mineral – fatty blends, and sulfurized mineral – fatty blends. The 'active' fluids react chemically with the metal surface and are dark or transparent sulfurized oils, sulfachlorinated oils, and fatty-compounded sulfurized and sulfachlorinated oils.

Chemical and Synthetic Fluids These contain organic and/or inorganic materials and form solutions rather than emulsions in water. They are generally transparent and allow observation of the work. Common ingredients are rust

192

inhibitors, mild lubricants, germicides, wetting agents, and water conditioners. Synthetic fluids include polyalkylene glycols and silicones which have a wide temperature range of operation.

Gases These may be used to penetrate the space between the tool and workpiece and provide an inert atmosphere around the cutting zone. To improve cutting capabilities, the gases may be refrigerated and, when applied to the cutting zone, sublime and remove heat. Their major disadvantage is expense, although carbon dioxide, air, and nitrogen find some usage.

Applications

Cutting fluid applications are of two types:

1. those used for low speed cutting, where finish is of primary concern, and
2. those used for high speed cutting, where finish is secondary to tool life and cooling

Of the gaseous types used, carbon dioxide is particularly effective in reducing wear on carbide tools when machining titanium alloys, inconels, and other materials which are difficult to machine.

Chemical and synthetic cutting fluids have high heat conductance and are therefore suitable for severe machining operations such as reaming, tapping, threading, broaching, and sawing. They are also suitable for machining cast, nodular, and malleable irons and their use does not require workpiece degreasing. High carbon and alloy steels are often machined with chemically active oils containing sulfurized fatty oils or other compounds. If severe frictional problems exist, and lubrication, oiliness, and anti-welding properties are critical, oil type cutting fluids are preferred. For machining tough, stringy, low carbon steel, sulfurized mineral oils are used and the sulfur in the fluid forms a film on the metal surface which, because of low shear strength and brittleness, facilitates cutting. In machining tough alloy steels, especially in severe machining operations, extreme pressure additives such as sulfur and chlorine are added to cutting fluids.

Greases

The US National Lubricating Grease Institute defines greases as solid or semi-fluid lubricants consisting of a thickening agent in a liquid lubricant which may include other ingredients imparting special properties. Thickening is required so that a grease will remain in contact with metal surfaces without leaking under gravity or centrifugal action and will not be squeezed out under pressure. A grease must be able to flow when pumped in many applications. A grease should also reduce power required to operate machinery, particularly under starting conditions.

The most important ingredients used in grease manufacture are the oil and thickening agent or soap to be compounded. The soap is normally obtained from animal or vegetable fats mixed with an alkali, calcium and sodium being the most common. Lithium, however, is becoming more frequently used;

aluminum, lead, and barium may also be employed. The oil used in grease can be either petroleum or mineral oil. Any grease should contain the same type of oil that would be used if the lubrication application were by oil alone.

The type of oil used has a considerable effect on grease oxidation and rubber swelling. Other materials used in grease are solid thickeners, fillers, and additives, and include a variety of materials such as fatty acids, neutral fats, oxidation inhibitors, extreme pressure agents, rust inhibitors, wear prevention agents, dyes, stringiness agents, hydrogen suppressants, odorants, color stabilizers, metal deactivators, water repellants, viscosity index improvers, pour point depressants, foam enhancers or depressants, melting point improvers, and others. Graphite, molybdenum disulfide, talc, or metal powders are also used in special greases. With all additives or combinations of some of them, grease is produced in two main types: boiled greases and cold set greases.

Boiled greases are made by mixing soap materials together with lubricating oil. They are mixed in an autoclave and heated until the soap is foamed within the oil. The mixture is then transformed to an open vessel where more oil is added and the grease is stirred until cool. With lithium or aluminum based greases, the soap is dissolved and dispersed throughout the oil. Care is needed in cooling to control the final consistency. Cold set greases are made from soaps derived from lime and resinous oils. Part of the oil is mixed with resin and part is mixed separately with lime and water, the two mixtures then being combined when warmed. The process is very simple and cheap, but the grease quality is not comparable with that of boiled greases.

Properties

Lubricating greases are usually classified according to their consistency and the nature of their soap bases. The consistency may vary from hard block to liquid greases, the latter closely resembling ordinary oils. Other properties depend on the lubricating oil viscosity and the additives. The melting point of a grease, at which it loses its semi-solid character, is as important as is its resistance to water. Since the main function of grease is to lubricate bearings or rotating mechanisms, the breakdown of greases at high speed is also important. Lithium greases and some synthetics are best for such applications.

There are many instances where there is little to choose between the performance of a grease or an oil, but in general greases are used where it may be difficult or inconvenient to maintain a supply of oil. Greases are also better able to exclude moisture, dust, and corrosive gases, especially in ball and roller bearings. They are suited to use in slow moving, heavily loaded bearings with reasonably large clearances where the condition of hydrodynamic oil lubrication is unattainable. Greases will not withstand continual mechanical shear nor, with the exception of some lithium greases and synthetics, withstand temperatures much greater than 95 °C.

Types

Lime based greases are characterized by their smooth buttery texture. They are made up of a lubricating oil, a lime soap and sufficient water to produce some stability and are particularly valuable where lubricated parts are subjected

194

to washing. Their disadvantages are poor performance at low temperatures, they suffer dehydration (especially in extended storage), and poor long-term stability.

Soda based greases are noted for their fibrous or spongy texture. They emulsify in the presence of water and are impractical in applications where water is present. They can stand much higher temperatures and machinery speeds than lime based greases, and their resistance to oxidation in continuous service is good.

Lithium based greases are expensive, but possess some outstanding advantages in ball and roller bearing lubrication in that they function effectively at high speeds and high temperatures. They have high resistance to water, combined with excellent stability. Barium based greases have somewhat similar characteristics.

Lead based greases consist of a mixture of lime soap, mineral oil, and a lead-carrying additive which aids load carrying capacity.

Synthetic greases are expensive and are not made in large volume production. Fluorocarbon greases are produced by blending oils with PTFE wax and are inert to highly reactive fluids such as oxygen, nitric acid, and hydrogen peroxide. Synthetic greases made with silica gel cannot be used in contact with caustic solutions, hydrofluoric acid, or gaseous fluorides. The most notable feature of synthetic greases is their performance at elevated temperature (to 280 °C). They are also noted for high load carrying capacity.

Selection of Grease

The selection of a grease is based on its characteristics and the required operating conditions. Characteristics requiring attention are purity, consistency, chemical and thermal stability, and adherence. Bearing operating conditions include speed, temperature, load, and location. Heat conductance of most greases is poor and excess grease may retain heat.

Gear greases are normally sodium, calcium, or aluminum types, while greases for rolling contact usage may be of any type. For rough machinery and axle lubrication, greases of the cold-set variety are normally satisfactory. Block greases are shaped to fit into bearings running at high temperatures. They usually have a calcium or sodium base, are exceptionally hard, and have high melting points. Greases used in water pumps and other machinery where frequent contact with water is experienced are normally of the calcium type. Synthetic greases are used exclusively in special high temperature and high load carrying applications.

References

FREEMAN, P. *Lubrication and Friction*, Pitman, London, 1962
NEAL, M.J. *Tribology Handbook*, Butterworths, London, 1973
CAMERON, A. *Principles of Lubrication*, Longman, London, 1966
American Society of Tool and Manufacturing Engineers. *Tool Engineers Handbook*, McGraw-Hill, New York, 1959

11
Materials Testing

Destructive Methods

Originally, materials (or mechanical) testing was confined to the determination of mechanical properties under simple tensile or compressive loading. By the late 1950s the number of different mechanical tests was large. Whereas originally mechanical testing was concerned with the determination of the limits within which elastic theory could be safely applied to engineering structures, it has changed its character completely and has become largely a method of quality control. In fact, many mechanical tests are no longer concerned with mechanical properties in a strength of materials sense, but with the properties required in materials fabrication, such as steel sheet drawing and pressing or of steel plate welding.

There are many tests—both destructive and nondestructive—of which engineers should have at least some knowledge. A number of these are tabulated in Table 13. In this chapter only a few major types of destructive tests are described.

Tensile Test

The tensile test occupies a unique position among mechanical tests and many materials are assessed based on it. It is simple and economical, it takes little time, it does not require much material or elaborate apparatus other than the testing machine. No precision measuring equipment is required and the test can be carried out by semi-skilled or unskilled labor. The results can be graphically recorded. Tensile testing, as with almost all materials testing, is governed more by simplicity, convenience, cheapness, and expediency rather than by scientific requirements.

The tensile test is carried out with a prismatic or cylindrical piece of material of standardized dimensions. Its ends are clamped in the testing machine, which applies a tensile load, the magnitude of which is increased from zero to its maximum value at a specified rate. By plotting the load against the extension of the test piece measured over a specified length, the load – extension diagram is obtained.

Table 13   Types of Materials Tests

Abrasion
Adherence
Arc resistance
Bearing strength
Brittle fracture
Brittle point
Brittleness temperature
Burst strength
Cavitation erosion
Cleavage
Cohesive strength
Corrosion (stress, erosion)
Cold flow
Complex modulus
Compression fatigue
Crack growth resistance
Creep (strength, rate, recovery)
Cross-cut adhesion
Crushing strength
Damping/dissipation/ hysteresis
Delamination
Dielectric strength
Dissipation factor
Ductile fracture
Irradiation
Fatigue (strength,endurance limit, runout)
Fatigue notch factor
Flaking
Flash point
Galling
Glass transition temperature
Hardness
    Durometer
    Brinell
    File
    Knoop
    Rockwell
    Shore
    Vicat
    Vickers

Heat distortion temperature
Impact strength
Kink test
Knot test
Low temperature compression set
Melting point
Microhardness
Minimum bend radius
Modulus of rupture
Multiaxial
Outgassing
Peel strength
Permeability
Penetration
Proof test
Relaxation
Repeated blow impact
Residual elongation
Residence
Spalling
Splitting
Springback
Strain hardening
Strength
Stress corrosion
Stripping strength

Swelling
Tear resistance
Tenacity
Thermal stress, shock, fatigue
Yield, ultimate, wet strength

Compression Test

A compression test piece, generally in the form of a prism, cube, or cylinder, is loaded between the platens of a compression testing machine. Ductile materials such as mild steel deform elastically up to a certain limit followed by plastic deformation and the test piece, if the metal is ductile, forms a 'barrel' shape under increasing load. Brittle materials such as rock, brick, cast iron, and concrete exhibit much greater compressive strengths compared to their tensile strengths. The crushing strength of concrete determined by crushing a cube or cylinder may reach values of 35 MPa or more. Granite may reach 80 MPa and cast iron 200 – 500 MPa.

Notch Impact Test

There are various forms of the notch impact test. For the Charpy V-notch impact test, a specimen in the form of a simple supported beam approximately 50 mm long and 10 mm × 10 mm in cross-section, with a sharp 2 mm deep 45° V-notch machined in the center of the underside, is broken by a sharp blow from a swinging mass striking the specimen. The energy absorbed by the specimen in fracturing is measured.

There is no way of relating results obtained in this test to elasticity theory. Nevertheless, the test is most valuable, and it gives an indication of material toughness under conditions of impact or shock loading. Under such conditions, and in the presence of a notch producing stress concentration, ductile and deformable materials may break in a brittle manner without absorbing much energy. A lowering of testing temperature has the same effect and the notch impact test is considered a valuable test for steels used in welded structures and exposed to low service temperatures.

Hardness Tests

With increasing mechanization there has been an increasing demand for precise evaluation of hardness, particularly of metals. Many tests have been developed to meet the demand. In the Brinell hardness test, a load is applied to a hardened steel ball 10 mm in diameter which is in contact with a plane surface of the metal specimen. The load is maintained for 10 s or more and then removed. The diameter of the impression remaining is then measured and its spherical area calculated, assuming the same radius for the indentation and ball indenter. The Brinell hardness number is then obtained by dividing the load by the area. A limitation of the Brinell test is that its use is restricted to metals softer than the indenter. Metals used in tools frequently attain that level of hardness.

In about 1920 Rockwell introduced a test suitable for high hardness levels. It employs a diamond cone indenter having a spherical point. A hardness number is obtained from the difference between the depths of the two impressions obtained under a major load and a minor load. The number is read directly from a dial on the machine. Since fast and reliable readings are readily obtained, it is used widely. The hardness scale, however, is nonlinear relative to the Brinell scale and has a limited range of high sensitivity. A variety of scales is therefore required. These scales utilize a range of major loads and hardened steel indenters as well as diamond indenters. Low loads and large indenters are used to test relatively soft materials such as bearing metals and plastics.

The first hardness tester to meet acceptably the demand for a continuous sequence of comparable hardness values over the wide range of metals was the Vickers hardness tester. In this method, a square-based diamond pyramid indenter is used, having a 136° angle between opposite faces. The angle was chosen as the mean for the range of penetrations recommended with the Brinell test. At this angle, identical hardness numbers are obtained on relatively soft metals with both Vickers and Brinell tests. Loads from 100 to 1000 N are used in the Vickers test and the lengths of the impression diagonals are measured with an ocular micrometer. While this method is much preferred for laboratory work, its operation is rather slow for routine inspection.

There are many materials such as glass and minerals which will crack under diamond indenters. The Knoop hardness test overcomes this limitation by the use of an elongated diamond-shaped indenter. With this indenter, the hardness of extremely brittle materials, including glass and diamonds, can be measured without cracking or spalling.

Fatigue Tests

Many attempts have been made to deduce the strength of metals under the repeated application of loading, which is called the fatigue strength. Even in the late 1950s it was believed by some that the fatigue strength of a metal was a definite fraction of its static strength. This, unfortunately, is not the case, and fatigue strength must be determined by special testing. Fatigue testing machines are capable of applying many loading and unloading cycles rapidly, which enables tests to be completed within reasonable times. Several specimens are tested with different values of applied stress and loaded cyclically until failure. A diagram of stress versus the number of cycles to cause failure $(S-N)$ is obtained. The $S-N$ diagram tends—for certain materials, particularly steels— to run asymptotically to the N-axis as the stress is reduced. It is thereby concluded that there will be a maximum value of stress which could be applied infinitely often without producing failure; this is termed the endurance limit, the fatigue limit, or the 'run out' strength. For aluminum alloys, however, there is no definite fatigue limit.

Nondestructive Testing

Nondestructive testing (NDT) employs indirect measurements that do not damage structures or test objects in detecting flaws. They are used to evaluate the integrity of engineering materials, parts, assemblies, and structures, both during manufacture and in service. Test indications must be correlated with strength or serviceability determined by past experience or by destructive mechanical tests on similar items.

Human senses often serve as nondestructive tests. Visual inspection may reveal surface discontinuities and finish. Dimensions are gaged and surface irregularities may be detected by touch. Cracked metallic parts are detected by tapping and listening to their vibrations. For centuries, craftsmen have controlled the quality of their work through frequent inspections. Modern high speed production methods, however, offer little opportunity for such direct sensory inspection. In many manufacturing operation, parts move at high speeds behind shields and workmen observe machine performance rather than the products being fabricated.

Nondestructive tests are used to discover minute defects or hidden internal conditions that can cause premature service failures under severe operating conditions such as extreme temperatures, high stresses, and damaging environments. They are essential in prevention of disastrous failures of complex engineering systems such as those required for nuclear power systems, jet aircraft, missiles, space vehicles, and military equipment, and in process industries such as chemical and petroleum refining. They can also detect in-service damage or deterioration and are cheaper and faster than dismantling complex equipment for direct inspection. Economic benefits result from their use by eliminating waste of materials, loss of machine and labor time, and user dissatisfaction.

Probing Media and Detection Systems

Each nondestructive test method requires some form of probing medium to explore the test object and a detection system to reveal its reactions to discontinuities or material faults. Probing media include electromagnetic waves, mechanical vibrations, electric and magnetic fields, liquid penetrants, heat, movement of electrons or ions, and other forms of energy or motions of matter. Detector signals must be amplified to useful levels and converted to forms suitable for human interpretation or for actuation of display, recording, or control devices. Many different physical effects are used in probing media, detectors, and amplifiers. A few of the more important methods are next described.

Penetrating Radiation Tests

Electromagnetic waves similar to light waves but of much shorter wavelengths are used as probing media in tests based on electromagnetic radiation. Radiation sources include X-ray generators and radioisotopes that emit γ rays. Their high energy photons have wavelengths that are small compared with interatomic spacings and can penetrate solid materials which are opaque to light. Radiation is partially absorbed and scattered within materials, depending upon wavelength, material composition, thickness, and density. The transmitted radiation beam forms shadow images of voids and discontinuities, changes of material thickness, and areas of differing densities.

In film radiography, the shadow image is recorded on film. In direct fluoroscopy, a phosphor screen fluoresces with a brightness proportional to the radiation intensity. Electrostatic image tubes can be employed to enhance the image brightness. In xeroradiography, a photoconductive coating on a metal plate is

200

charged electrically and discharged locally when exposed to a radiation image. The resultant latent electrostatic image is developed by spraying the exposed plate with oppositely charged pigmented particles.

Special X-ray sensing television camera tubes can be used in closed circuit television systems that reproduce images at a distance from the radiation source. Human observers interpret the X-ray images to detect internal discontinuities such as gas holes, porosity, slag inclusions, shrinkage defects, cracks, and other flaws in castings, weldments, and brazed bonds. Positions and dimensions of internal components in complex assemblies are also measured from X-ray images to quantify loss of thickness due to corrosion, fatigue cracking, and other forms of deterioration.

Geiger counters, scintillation counters, ionization gages, semiconductor crystals such as cadmium sulfide and other detectors provide signals proportional to radiation intensity. They can be used in thickness or mass-per-unit-area gages to measure sheet materials in rolling mills or coatings applied to sheet materials. They are often used with feedback controls to reduce process variations. X-ray diffraction and fluorescence analysis techniques are used to identify constituents, measure residual strain, and follow structural changes during processing.

Ultrasonic Tests

Sound waves or mechanical vibrations at high frequencies (from 200 kHz to 25 MHz) are used as the probing media for ultrasonic nondestructive tests. They are created in piezoelectric transducers are transmitted into solid test materials through films or layers of liquid couplants such as oil or water. Ultrasonic beams can be directed or focused similarly to light beams. They are partially reflected at voids, cracks, and interfaces where there are changes in material density or elasticity, and at interfaces between materials in which sound travels at different velocities.

Ultrasonic waves propogate well in fine-grained wrought materials, but tend to be attenuated and scatter in coarse-grained structures such as large castings. The echoes which return from material boundaries or discontinuities can be used to measure material thickness, to detect discontinuities, and to indicate some material properties. Under ideal test conditions, echoes from very small flaws can be detected through considerable thicknesses.

Many variations exist in ultrasonic test methods. In ultrasonic resonance testing, longitudinal waves of varying frequencies establish resonance in parallel surfaced parts when the round-trip transit time equals the time between successive waves from the transducer. This permits measurement of material thickness, detection of laminar flaws in sheets or plates, and detection of wall thinnings due to wear or corrosion. In pulse rejection testing, a brief burst of sound waves travels from transducer to test object, reflecting from the opposite surfaces or from intervening discontinuities. The initial pulse and return echoes are displayed on a cathode ray oscilloscope as a function of time, so that the flaw depth or part thickness can be measured. Echo magnitudes can be used to estimate the areas in comparison with standard test blocks containing drilled holes. The method provides extreme sensitivity to cracks and flaws, voids, inclusions, segregations, and other discontinuities.

In contact ultrasonic tests, the transducer is held manually in contact with test object surfaces using an oil film couplant. In other tests, the object is immersed in water and the transducer is supported at a distance from the surface being scanned. Large-scale automatic systems have been developed to display cross-sectional or map-like facsimile images of test objects and internal discontinuities. Such systems are used to inspect shells and plates, tubes, forgings, and complex assemblies such as bonded laminates or brazed honeycomb structures with excellent resolution of detail.

Electromagnetic Induction Tests

An alternating magnetic field is used as the probing medium in electromagnetic induction tests. The test object is placed within the influence of a magnetized coil carrying alternating current of suitable frequency. Eddy current and varying states of magnetization are induced within the object by transformer action. The magnetic fields of the test object are superimposed upon the exciting field and can be detected either through their reaction on the exciting coils or by means of an additional pick-up system. The amplitude and phase relations of the output signals are analyzed by electronic circuitry to separate effects of test material conductivity, permeability, dimensions, shape, and flaws.

Rod, bar, tubes, and symmetric test objects can be passed through test coils at high speed to measure dimensions, alloy composition, hardness, and other characteristics or to detect discontinuities such as cracks, seams, laps, wall thinning, and others. Small probe coils can also be scanned over test parts to measure local discontinuities or property variations. Eddy current tests are suited to high speed automatic production processes, since measurements often take only a few thousandths of a second. They produce electrical signals which can be analyzed statistically by electronic circuits and used directly to control processes or to produce records.

Magnetic Field Tests

Ferromagnetic materials such as steel and iron can be tested by using magnetic fields as the probing medium. In magnetic particle tests, the parts are magnetized and a suspension of finely divided particles is applied to the surface. Where surface or near surface discontinuities transverse to the direction of magnetization introduce gaps in the magnetic flux paths, some of the flux lines tend to 'leak out' at the surface, attracting accumulations of particles which delineate discontinuities. Visible or fluorescent dyes applied to the particles permit them to be readily seen, even when the discontinuities are invisible because of their minute size or subsurface location. The test finds wide application in detection of cracks, seams, laps, inclusions, and segregations in ferrous materials.

202

In magnetic probe tests, a sensitive magnetometer or 'Hall effect' detector is used to measure magnetic fields at the surfaces of ferromagnetic materials. Measurements of the residual field or its coercive force permit evaluation of material hardness, strength, or heat treatment. Ferromagnetic material thickness can be measured in terms of saturation flux linkages. Anisotropy in rolled sheet materials can also be detected.

Liquid Penetrant Tests

The probing medium used in liquid penetrant tests consists of a light petroleum distillate containing colored or fluorescent dyes. After cleaning, test parts are immersed in or sprayed with liquid penetrant which enters surface-connected discontinuities such as hairline cracks. After penetration, excess surface fluid is washed off and a porous developer coating which serves as a blotter is applied. Penetrant trapped in discontinuities then seeps back into the developer coating, tending to spread laterally so as to amplify flaw indications.

An alternative system involves the application of an emulsifying agent to aid in removal of the surface penetrant, and provides increased sensitivity. Surface-connected seams, cracks, laps, and other discontinuities are shown by streaks or dots when examined under suitable illumination. Test sensitivity is adequate to reveal discontinuities too fine to be seen with a microscope, since wavelengths of visible light are too great to permit light to enter fine defects.

Filtered Particle Tests

Porous materials such as ceramics may be inspected by means of solid particles suspended in a liquid penetrant. When the liquid penetrant soaks into the porous material, the particles are filtered upon its surface. At cracks and other surface connected discontinuities more penetrant enters and numerous particles are filtered out at the entrance. Accumulations of particles coated with visible or fluorescent dyes may be readily seen. Wet clay materials are inspected in this manner prior to kiln firing, since defects present at this stage can be corrected at low cost.

Thermal Tests

Heat is used as the probing medium in thermal tests of metallic materials and bonded assemblies with the resultant temperature distributions being indicated by temperature-sensitive coatings or remote infrared detectors. Temperature-sensitive paints change color at specific temperatures and temperature-sensitive phosphors vary in fluorescent brilliance under ultraviolet excitation in response to gradients in temperature. Heat-repelled fluid coatings are applied by spraying and test parts pass under infrared heat lamps which create temperature gradients and evaporate the coating solvent. Test objects emerge with a varnish-like coating on which are observed temperature distributions attained during the heat treatment.

Other Methods

Many other nondestructive tests have been developed. Electric current tests are used to detect transverse fractures in train rails. Electrified particles provide probing electric fields and detectors for discontinuities in insulating materials and coatings. Brittle lacquers and birefringent plastics are used in coatings to detect strain distributions. Pressure and leak tests may be performed with liquid penetrants and gases such as helium or hydrogen, which can be detected in minute concentrations by mass spectrometry. Chemical reagents which react with specific constituents of test materials are employed in chemical spot tests.

Test Limitations and Operator Requirements

All nondestructive tests are specific and have limited applications and capabilities. Complete inspection and quality assurance usually requires use of two or more tests to assess more fully material conditions and discontinuities that could lead to premature failure in service. All nondestructive test methods are indirect and their indications must be correlated with serviceability by other means, such as destructive tests of similar specimens under simulated service conditions or by extensive service experience with similar objects or materials. Frequency of inspections and sampling ratios are of prime importance in using nondestructive testing.

In most test methods which involve human interpretation, training and judgement of the inspector are vital to test reliability. The inspector must have adequate knowledge of the nature of materials and their processing, of design requirements and conditions of service, and of the influence of discontinuities upon performance. Such extensive knowledge is rarely available and continued indoctrination and training are essential in maintaining inspector qualification.

Suggested Further Topics

It is hoped that the reader of this chapter on materials testing will study the further topics suggested below. Additionally, other teaching institutions which may use the present book as a text or reference might wish to use the topics listed for student seminars or research projects:

abrasion and wear
corrosion
fracture toughness
high strain rate effects
high temperature effects
irradiation effects
low temperature effects
stress concentration
thermal shock and fatigue

Laboratory sessions dealing with quantitative identification of real materials are particularly valuable in the study of engineering materials. In the past, entire classes of the author's have undertaken identification of parts and components ranging from commonplace vacuum cleaners and washing machines to fairly exotic aircraft flight instruments, such as a gyrocompass and a variable-pitch propeller drive mechanism. Display boards of the student-identified materials and data are of great interest and the process involved in positive identification is often challenged by peer groups.

12
Computerized Materials Selection

The selection of engineering materials is often an extremely difficult task. The practising engineer or designer tends either to become specialized in a given material's field or he tends to use or specify familiar materials, possibly being unaware of recent developments. The student engineer is usually completely bewildered by the hundreds of materials mentioned throughout his formal training. The difficulty in selecting materials is occasioned because no one individual can possibly possess in-depth knowledge of all types. There are always nagging questions: 'Is this the best metal or alloy to use?' 'I wonder if plastics or composites might do the job instead?' 'Which would be the best finish or coating to use here?' 'Would a cheaper material have sufficed?'

Consider, for instance, the breakdown given in Table 14 for various material classes and the number of subclasses (alloys, types, grades, various manufacturers, etc.) as presented in a recent *Materials Selector* published by Reinhold. With an approximate minimum of 20 properties for each material, there are therefore over 30 000 pieces of data with which to cope. Further, the data are given for representative materials only and do not reflect, for instance, all possible heat treatments, composition variations, arbitrary combinations of materials, or a wide variety of material supply sources. Data on adhesives *per se* are not included. Nonetheless, the reference is most useful for preliminary material selections in many applications.

Table 14   Quantities of Engineering Materials

Class of material	Number of variations or subclasses
Irons and steels	361
Nonferrous metals	374
Plastics	197
Composites	153
Rubbers and elastomers	41
Ceramics, glass, carbon, and mica	64
Fibers, felts, woods, and paper	148
Finishes and coating	218
Total	1556

The question at hand is 'how to make the most effective use of such volumes of such property data?' Allowing factors of $10^2 - 10^3$ for other data sources, in all, there may be between 3×10^6 and 3×10^7 pieces of engineering material data! Also, how can one introduce judgement or weighting factors for particular properties so as to select an optimum material for a given application? The scope of this chapter addresses these questions. Two conceptual and mathematical approaches are described briefly. Comparative results from a computer program developed specifically to deal with selection of plastics are given by Hanley and Hobson (ASME paper No. 73-Mat-CC, October 1973). Extension of the program to deal with other materials is also discussed.

Selection Criteria

With the large number of materials available, it is obvious that the selection for a given application can usually be made according to a broad class, since these have distinctive characteristics. For instance, unreinforced plastics rarely have tensile strengths greater than about 140 MPa and specific gravities are usually less than 2.0; their coefficients of thermal expansion are also much greater than those of metals. Such combinations of properties delineate plastics from metals.

Similarly, in comparison with other classes of materials, ceramics have unique characteristics of higher strengths in compression than in tension and distinctive thermal and electrical properties. Advanced composites also have particular combinations of anisotropy, strength (both static and fatigue), stiffness, and lower density compared with other materials. Other combinations of parameters may be used to differentiate fibers, rubbers, timbers, and so on.

The difficulty in a particular material selection is choosing one of the many types of subclasses from within a given class. Here there are no hard and fast rules and there is considerable scope for individual approaches. The two selection criteria described herein—termed 'geometrical' and 'algebraic—were developed as independent approaches to the problem. There are some similarites between the methods, e.g., both use numerical deviations and weighting coefficients, but they are sufficiently distinctive to warrant separate presentation. Both are incorporated as alternatives in a single minicomputer program.

Geometrical Approach

An ideal material for a specific application may be geometrically represented in the form of a regular polygon with various properties plotted along radials from the polygen center to each vertex. Five properties forming a pentagon might be strength, electrical conductivity, corrosion resistance, density, and cost. Values for each property are thus defined as Y_1, Y_2, ..., Y_n. These may be represented as equal distance radials. A candidate material for the application may then be considered with its respective properties—designated X_1, X_2, ..., X_n—plotted along each radial. Clearly, the suitability of a candidate material can be rated according to three factors: (1) the size of the polygon and its closeness to the ideal, (2) the shape of the polygon and its closeness to the ideal, and (3) the subjective assessment of the importance of deviations from the ideal.

Any number of properties might be represented on such polygons, including, say, moduli of elasticity, hardnesses, thermal conductivities, expansion coefficients, magnetic properties, corrosion rates for various environments, etc. Generally these are all positive quantities. The multidimensional problem may be reduced to only two dimensions as follows. A 'mean weighted characteristic' (MWC) may be defined by

$$MWC = \frac{\sum\limits_{i=1}^{n} \dfrac{\alpha_i X_i}{Y_i}}{\sum\limits_{i=1}^{n} \alpha_i}$$

where n is the number of properties and α_i is a weighting coefficient arbitrarily chosen from 0 to 1 according to the relative importance of a particular property, zero being unimportant and unity being a criticial property.

The computer can then rapidly calculate an MWC for all materials supplied to a data bank. The closer the MWC is to unity, the closer the overall properties meet the requirements, i.e., the polygons are nearly the same size. This is true when there are a sufficiently large number of properties of interest. However, in limiting cases, when many α values are set equal to zero, the concept is not strictly true.

Likewise, a 'balance factor' (BF) may be defined as a root mean square deviation given by

$$BF = \sum\limits_{i=1}^{n} \sqrt{\left(\frac{X_i}{Y_i} - MWC\right)^2}$$

The closer the BF is to zero, the nearer the shape of the candidate material polygon to that desired. The rating procedure or criterion for a materials selection is to choose those having minimum values of

$$\sqrt{(1 - MWC)^2 + (BF)^2} = d$$

The above equation represents a distance (d) on a plot of MWC versus BF data for various materials. It is possible to rate materials on a MWC – BF plot according to good, fair, or poor overall characteristics and according to a good or poor balance of characteristics. Initial screening of candidate materials discards those with MWCs < 0.8 and BFs > 0.2, saving computer time; those values being selected on the basis of experience with the program.

The computer, of course, does not draw polygons or plot MWC versus BF values—although it could—rather, it can be made to select a number of materials that are closest to the ideal and present them for consideration. Used in an interactive mode, the approach can also assist in evaluation of trade-offs, e.g., by changing emphasis on the importance of one property over another (cost, strength, etc.). Additionally, the approach may yield information for basic materials research—when, say the output reads 'NO SUCH MATERIAL EXISTS!' Alternatively, in such cases, limits on MWC and BF values may be relaxed to find a possible candidate.

Algebraic Approach

A very direct approach to choosing materials from a data bank is to base the selection process on minimizing the sum of the per unit deviations of the properties of candidate materials from the specified properties. If the specified properties are designated Y_i, the properties of candidate materials are X_i, and the weighting coefficients are α_i as before, then the criterion is expressed algebraically as follows:

$$\text{Min } Z = \sum_{i=1}^{n} \left| \alpha_i \frac{X_i}{Y_i} - 1 \right|$$

The algebraic approach to the materials selection problem is based on the above simple algorithm. However, the use of interactive computer terminals permits significant improvements without adding greatly to the demands on the program user.

In materials selection, maximum and minimum property values are specified frequently, but deviations above and below those values respectively can be tolerated. For example, a specified strength of 140 MPa is a 'soft' lower limit and a material with a strength of, say, 190 MPa should not be penalized for deviating significantly from the target. In special cases, such as for a specified failure strength or a desired thermal expansion, it may be desirable to penalize a deviation in any direction. The algebraic approach recognizes this range of possibilities and through the computer terminal asks the user for each specified property whether the limit is upper, lower, or target. The algorithm above incorporates these directional limits by introducing the following constraints:

$$\frac{X_i}{Y_i} > 1 \text{ for upper limit on property } i$$

$$\frac{X_i}{Y_i} < 1 \text{ for lower limit on property } i$$

$$\frac{X_i}{Y_i} = 1 \text{ for target property } i$$

When a limit is not violated there is no contribution to Z.

Occasionally it happens that a limit is fixed, for example an absolute upper limit on cost. The weighting coefficient could be used to transform any of the above limits from 'soft' to 'hard'. A value of α_i of 1000 would be sufficient to ensure that no material outside the limit is selected. Varying α_i in an interactive mode permits evaluation of trade-offs between different properties.

For any data bank, it is necessary to limit the number of candidate materials presented for consideration. Otherwise, the selection procedure will list most or all of the materials in store. In the algebraic approach, using a data bank of 31 materials, the selection process ignores materials if any property goes outside limits by a weighted value of 0.4 or if the sum of all weighted deviations exceeds 2.0. In symbolic form, materials are listed if and only if:

$$\alpha_i \left| \frac{X_i}{Y_i} - 1 \right| < 0.4$$

for all properties i outside limits, and $Z < 2.0$.

A conversational language such as BASIC permits limits on property deviations to be built into the program and yet be available for modification if desired. Many variatons on the algebraic approach are possible. Consideration has been given to incorporating nonlinear penalties on deviations from limits as, for example, in least square curve-fitting. Other penalty functions are possible but, based on experience gained with linear penalties, there appears to be no strong justification for any greater degree of sophistication.

Conclusion

Relatively simple numerical procedures are available for computer-aided selection of materials. The approaches have general applicability for many classes of materials. Feasibility of the approaches has been demonstrated specifically for a wide range of plastics on a PDP 8/e minicomputer. Data banks have also been developed for selection of ferrous and nonferrous alloys. Parameters in addition to mechanical properties (yield strength, hardness, modulus, and density) are cost, forms available, fabricability, and corrosion resistance.

The program operates efficiently in an interactive mode. It is especially useful as an educational aid and may readily be employed, updated, extended, or modified for industrial application. The approach has been found more direct and more readily implemented than some large-scale data retrieval software packages.

The quantitative process of materials selection as described above has been extensively and very recently applied by M. Farag in *Materials and Process Selection in Engineering*, Applied Science, London, 1979.

210

Appendix 1
Guide for Literature Searches

This is a brief guide designed to help the reader search for library material on topics covered in this book. It was prepared by the Library of the Capricornia Institute of Advanced Education, and is reproduced here by permission.

If you need help beyond the guidelines given below, you are advised to consult a suitable member of the staff of your library.

START WITH | REFERENCE BOOKS Dictionaries, encyclopedia and handbooks will help you define your question, perhaps answer it and certainly lead you to other helpful information.

R
603
MAC

McGraw-Hill Encyclopedia of Science and Technology, 1966, 15 vols

R
660.03
ENC

Encyclopedia of Chemical Technology, 2nd edn, 1963, 22 vols

R

660
CHE

Chemical Technology; an Encyclopedic Treatment, 1968, 7 vols

R
621.03
NAY

NAYLER, J. L. and NAYLER, G.H.F. *Dictionary of Mechanical Engineering*, 1967

R
624
CIV

Civil Engineer's Reference Book, 3rd edn, 1975

R
621.3
SAY

Electrical Engineer's Reference Book, 13th edn, 1973

R
621
PAR

Mechanical Engineer's Reference Book, 1973

R 620.16 AME	American Society of Mechanical Engineers. *Metals Engineering—Design—Process—Properties*, 1954 – 65, 3 vols, (ASME Handbooks)
R 620.00212 BOL	BOLZ, R. E. and TUVE, G. L. (eds). *Handbook of Tables for Applied Engineering Science*, 1973
R 620.11 MAT	*Materials Selector*, latest edn
R 668.4 AUS	*Australian Plastics and Rubber Buyers' Guide*, 1975
R 668.40212 GUI	*Guide to Plastics; Property and Specification Charts*, 1975
R 669 AME	American Society for Metals. *Metals Handbook*, 8th edn, 1961 + , 9 vols
R 669.003 CAG	Cagnacci-Schwicker, A. *International Dictorary of Metallurgy, Mineralogy, Geology*, 1968
R 671 AUS	The Australian Institute of Metals. *Handbook*, 2nd edn, 1973
R 671.3603 SIM	SIMONS, E. N. *A Dictionary of Metal Heat-treatment*, 1974
R 674.834 PLY	Plywood Association of Australia. *Plywood Manual*, 1972
R 678.2 CRA	CRAIG, A. S. *Dictionary of Rubber Technology*, 1969
R 674.132 PRI	Princes Risborough Laboratory. *Handbook of Hard woods*, 2nd edn, 1972

You will find other useful reference books through the subject catalog or by browsing the reference shelves.

212

CONSULT   CATALOG Check author/title catalogue if you know
                 the author's name or the title of a book or periodical. Use
                 the subject catalog when you want books on a particular
                 subject. Some headings which might help you search are

Alloys   Lubrication and lubricants
Aluminum   Machine tools
Aluminum alloys   Machinery
Asphalt   Materials
Bridges—Design   Mechanical engineering
Bridges—Specification   Metallurgy
Bridges, Concrete   Metals
Bridges, Iron and steel   Metals—Fatigue
Bulk solids flow   Metals—Fracture
Bulk solids handling   Metals—Testing
Carbon composites   Piling (Civil engineering)
Ceramic materials   Pipe lines
Chemical engineering   Plasticity
Civil engineering   Plastics
Composite materials   Plywood
Concrete   Polymers and polymerization
Concrete beams   Precast concrete
Concrete construction   Prestressed concrete
Elastic plates and shells   Protective coatings
Elastomers   Reinforced plastics
Engineering   Rock mechanics
Engineering design   Shells (Engineering)
Foundations   Steel
Fracture of solids   Strains and stresses
Glass reinforced plastics   Strength of materials
Grinding and polishing   Structural design
Heat engineering   Structural frames
Heat resistant alloys   Structures, Theory of
Industrial engineering   Timber
Iron founding   Vacuum technology
Iron—Metallurgy   Wood
Low temperature engineering   Zinc

FIND ADDITIONAL   INDEXES The following publications index
INFORMATION IN   articles from many periodicals. Use them to find
                      the most recent information and sometimes, the
                      only information on a 'new' topic. Ask a mem-
                      ber of the library staff if you need help using
                      these indexes.

R   *Engineering Index*, annual
016.62
ENG

R 016.6 A P P	*Applied Science and Technology Index*, monthly except July
R 016.6 BRI	*British Technology Index*, monthly
R 016.6201 APP	*Applied Mechanics Reviews*, monthly
R 016.621 ISM	*Ismec Bulletin*, twice monthly

Use the author/title catalog to find whether your library has a particular periodical.

Remember: ASK FOR HELP when you need it.

Appendix 2

Typical Properties of Engineering Materials*

Material	Tensile strength (MPa)	Modulus (GPa)	Elongation (%)	Poisson's ratio	Specific gravity
Acrylic	55	3	5	0.36	1.20
Aluminum					
Pure, annealed	83	69	40	0.34	2.71
1100 (BHN 26)	110	69	10		2.71
2014 T6 (BHN 155)	510	72	13		2.77
2024 T4	475	72	20		
7075 T6	580	71	11		
A108 (cast)	190	71	2		
Beryllium	480	290	3	0.10	1.86
Boron	3500	414		0.20	2.33
Brass 6A	240	97	40	0.35	8.47
Bronze 385	410	97	30		8.47
Cast iron					
30 (gray)	210	103	Nil	0.20	6.92
60 (gray)	410	138	Nil	0.25	7.48
Concrete (compression)	20	17		0.15	2.41
Copper					
Pure, annealed	230	117	50	0.36	8.94
Epoxy					
Cast, rigid	70	3	4	0.30	1.16
Glass					
Plate, soda lime	70	69	Nil	0.24	4.15

* At room temperature and normal loading rates.

Material	Tensile strength (MPa)	Modulus (GPa)	Elongation (%)	Poisson's ratio	Specific gravity
Graphite					
General purpose	7	7			1.66
Lead	17	14	45		11.4
Magnesium					
(wrought)	280	45	10		1.80
Molydenum	650	324			10.2
Nickel (annealed)	380	207			8.89
Nylon 6 (cast)	90	4	20	0.40	1.11
Polyethylene (medium density)	14	0.5	200		0.91
Polystyrene (molded)	35	3	2 – 30		1.08
Polyvinyl chloride (rigid)	40	3	5 – 25		1.38
Rubber					
Natural	20		800	0.50	0.94
Neoprene	20		850	0.50	1.23
Steels					
1005 (cold rolled, BHN 125)	450	207	30	0.27	7.84
1005 (hot rolled, BHN 90)	350	207	40	0.27	
1020 (normalized)	450	207	30	0.27	
1045 (BHN 450)	1600	207		0.27	
4130 (BHN 365)	1400	200		0.28	
4340 (BHN 410)	1500	200		0.30	
5210 (BHN 520)	2000	207		0.29	
9262 (BHN 410)	1600	200		0.27	
304 stainless (BHN 160)	740	186		0.27	8.03
350 stainless (BHN 496)	1900	179		0.30	7.81
18 Ni maraging	1900	186			8.03

216

Material	Tensile strength (MPa)	Modulus (GPa)	Elongation (%)	Poisson's ratio	Specific gravity
Teflon	20	0.4	300		2.16
Titanium	480	103	25		4.51
Tungsten	1500	379	Nil		19.4
Wood					
Australian softwood	80	10			0.51
Australian hardwood	140	19			0.99
Douglas fir	55	14			0.55
Oak	48	12			0.69